Praise for *The Rise of Wolf 8: Witnessing the Triumph of Yellowstone's Underdog*

"[Rick McIntyre's] greatest strength is the quiet respect and wonder with which he regards his subjects, a quality clearly informed by decades of careful watching."

PUBLISHERS WEEKLY

"Rick's book [*The Rise of Wolf 8*] is a goldmine for information on all aspects of wolf behavior and clearly shows they are clever, smart, and emotional beings."

PSYCHOLOGY TODAY

"The main attraction of this book, though, is the storytelling about individual wolves, including the powerful origin story of one of Yellowstone's greatest and most famous wolves."

WASHINGTON POST

"Yellowstone's resident wolf guru Rick McIntyre has been many things to many people: an expert tracker for the park's biologists, an indefatigable roadside interpreter for visitors, and an invaluable consultant to countless chronicles of the park's wolves—including my own. But he is first and foremost a story-teller whose encyclopedic knowledge of Yellowstone's wolf reintroduction project—now in its 25th year—is unparalleled."

NATE BLAKESLEE, *New York Times* bestselling author of *American Wolf*

"For many years I've thought that Rick McIntyre is the 'go-to-guy' for all things wolf, and his latest book, *The Rise of Wolf 8*, amply confirms my belief. A must read—to which I'll return many times—for anyone interested in wolves and other nature. Wolves and humans are lucky to have Rick McIntyre."

MARC BEKOFF, PhD, author of *Rewilding Our Hearts* and *Canine Confidential*

Praise for *The Reign of Wolf 21: The Saga of Yellowstone's Legendary Druid Pack*

———

"Like Thomas McNamee, David Mech, Barry Lopez, and other literary naturalists with an interest in wolf behavior, McIntyre writes with both elegance and flair, making complex biology and ethology a pleasure to read. Fans of wild wolves will eat this one up."

KIRKUS starred review

"Rick's passion for the Yellowstone wolves flows through this meticulous book about wolf love, play, life, and death. It's just like being there."

DR. DIANE BOYD, wolf biologist

"Rick McIntyre is a master storyteller and has dedicated his life to wolves—most particularly Yellowstone wolves. He tells their stories better than anyone, arguably better than anyone in history. I too have dedicated my life to wolves, yet reading Rick's stories, I still learn new things. This book is a treasure."

DOUGLAS W. SMITH, senior wildlife biologist and project leader for the Yellowstone Gray Wolf Restoration Project

"I'm always eager for the next book by Rick McIntyre. I learn so much fascinating information about wolves and their interactions with each other and with their prey."

L. DAVID MECH, author of *The Wolf: The Ecology and Behavior of an Endangered Species*

"Rick McIntyre has observed wild wolves more than any person ever. It is the way he sees wolves—as fellow social beings with stories to share—that makes his books so powerful. Through that lens, we glimpse our own hopes and dreams."

ED BANGS, former U.S. Fish and Wildlife Service wolf recovery coordinator for the Northern Rockies

RICK MCINTYRE

THE
Redemption of Wolf 302

FROM RENEGADE TO YELLOWSTONE ALPHA MALE

GREYSTONE BOOKS
Vancouver/Berkeley/London

21 22 23 24 25   5 4 3 2 1

Greystone Books Ltd.
greystonebooks.com

Cataloguing data available from Library and Archives Canada
ISBN 978-1-77164-527-0 (cloth)
ISBN 978-1-77164-528-7 (epub)

Editing by Jane Billinghurst
Copyediting by Rhonda Kronyk
Proofreading by Meg Yamamoto

Maps by Kira Cassidy
Jacket and text design by Fiona Siu
Jacket photograph by Doug Dance. This shot of one of 302's daughters shows how
302's charisma and striking looks were passed down through the generations.

Printed and bound in Canada on FSC® certified paper at Friesens. The FSC® label
means that materials used for the product have been responsibly sourced.

Greystone Books gratefully acknowledges the Musqueam, Squamish, and
Tsleil-Waututh peoples on whose land our Vancouver head office is located.

This book was written after the author finished working for the National
Park Service. Nothing in the writing is intended or should be interpreted as
expressing or representing the official policy or positions of the US
government or any government departments or agencies.

Greystone Books thanks the Canada Council for the Arts, the British
Columbia Arts Council, the Province of British Columbia through the
Book Publishing Tax Credit, and the Government of Canada for
supporting our publishing activities.

Canada

CONTENTS

"*If a man does not keep pace with his companions,*
perhaps it is because he hears a different drummer."

HENRY DAVID THOREAU, *WALDEN* (1854)

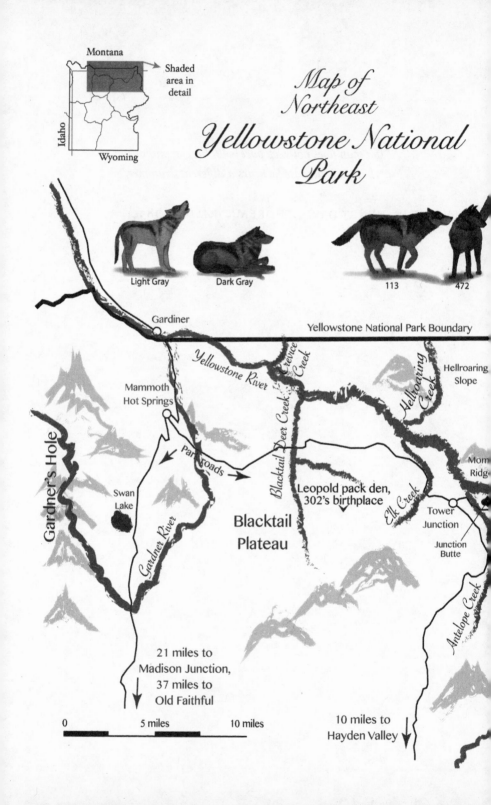

Map of
Northeast
Yellowstone National
Park

Montana

Idaho

Wyoming

Shaded area in detail

Light Gray   Dark Gray   113   472

Gardiner

Yellowstone National Park Boundary

Yellowstone River

Trevice Creek

Hellroaring Slope

Hellroaring Creek

Mammoth Hot Springs

Blacktail Deer Creek

Park roads

Leopold pack den, 302's birthplace

Elk Creek

Mom Ridge

Gardner's Hole

Swan Lake

Gardner River

Blacktail Plateau

Tower Junction

Junction Butte

21 miles to Madison Junction, 37 miles to Old Faithful

0   5 miles   10 miles

Antelope Creek

10 miles to Hayden Valley

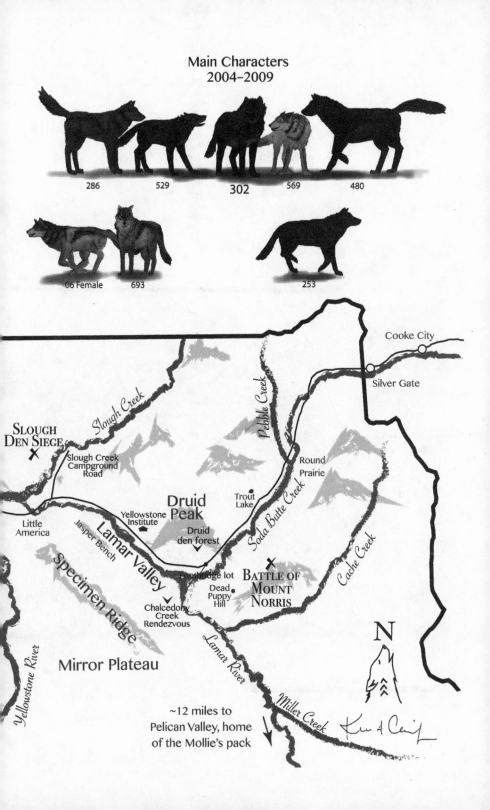

Main Characters
2004–2009

286   529   302   569   480

06 Female   693   253

Cooke City

Silver Gate

Slough Creek

Pebble Creek

SLOUGH
DEN SIEGE ✕

Slough Creek
Campground
Road

Round
Prairie

Druid Peak

Trout
Lake

Soda Butte Creek

Yellowstone
Institute

Little
America

Druid
den forest

Jasper Bench

Lamar Valley

Footbridge lot

Dead
Puppy
Hill

BATTLE OF
MOUNT
NORRIS ✕

Cache Creek

Specimen Ridge

Chalcedony
Creek
Rendezvous

N

Mirror Plateau

Lamar River

Yellowstone River

Miller Creek

~12 miles to
Pelican Valley, home
of the Mollie's pack

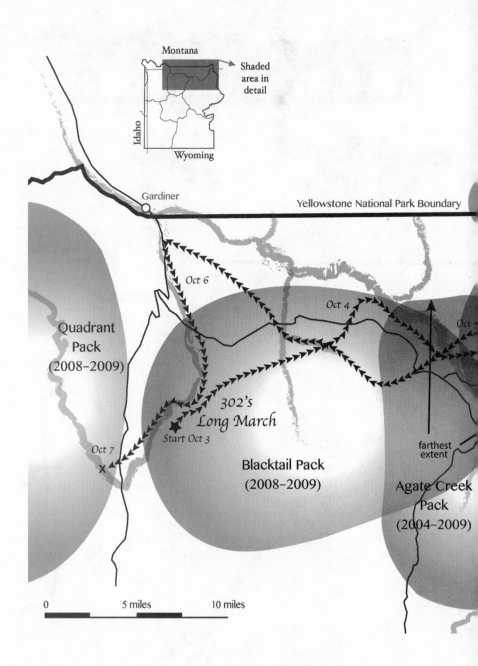

Montana

Shaded area in detail

Idaho

Wyoming

Gardiner

Yellowstone National Park Boundary

Oct 6

Oct 4

Oct

Quadrant Pack (2008–2009)

302's Long March

Oct 7

Start Oct 3

X

farthest extent

Blacktail Pack (2008–2009)

Agate Creek Pack (2004–2009)

0   5 miles   10 miles

Select
Yellowstone Wolf Pack
Territories
2004–2009

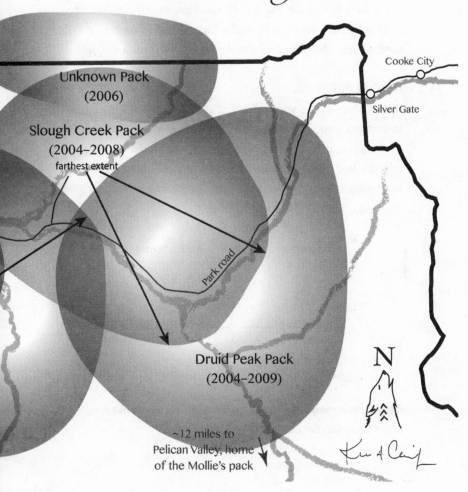

Cooke City

Unknown Pack
(2006)

Silver Gate

Slough Creek Pack
(2004–2008)
farthest extent

Park road

Druid Peak Pack
(2004–2009)

N

~12 miles to
Pelican Valley, home
of the Mollie's pack

PRINCIPAL WOLVES

THIS IS A list of the main packs and pack members covered in this book. Originating packs for individual wolves are in parentheses. M indicates a male and F indicates a female.

Druid Peak Pack

In early 2004, the alpha pair was 21M and 42F. 42 died early in the year, and 21 died a few months later. After the death of the alpha pair, the second-ranking male in the Druid pack, 253, looked ready to take over as the alpha male, but then two wolves from the Leopold pack, 302 and his nephew 480, showed up, and everything changed.

Alpha male		Alpha females
		286F to 2005
480M (Leopold)	—○—	**529F** to 2006
		569F

Other pack members mentioned by name or number

302M (Leopold) beta male	375F
255F	376F

570M	Dull Bar (F)
571F	Big Brown (M)
High Sides (F)	Medium Gray (M)
Low Sides (F)	Big Blaze (M)
White Line (F)	Small Blaze (M)
Bright Bar (F)	

Slough Creek Pack

In early 2004, former Druid 380F took over as the alpha female and, after the death of alpha male 261 in mid-2004, paired up with a big black male from the Mollie's pack, the first of three alpha males she would pair with over the course of this story. When she died in September 2008, Hook became the next alpha female. After 590M left the pack in late 2008, Hook paired up with 383M.

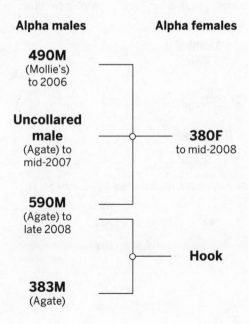

Other pack members mentioned by name or number

377M beta male (Mollie's)	527F beta female
629M beta male (Druid)	526F
489M third-ranking male (Mollie's)	Sharp Right (F)
Slight Right (M)	Slant (F)

Agate Creek Pack

The alpha pair for most of this story was 113M from the Chief Joseph pack and 472F, a former Druid. When 113 became too old to lead the pack, his son, 383, stepped up into the role. 383 ended up leaving the pack for the Sloughs because he and 472 were too closely related to mate, and Big Blaze, a wolf from the Druid pack, took over as alpha male.

Alpha males	Alpha female

113M
to 2007

383M ———○——— **472F**
to 2008   (Druid)

Big Blaze
(Druid)

Other pack members mentioned by number

471F	693F
642F	06F
692F	07F

Outsider Males

Two brothers from the Mollie's pack, 378 and 379, who sometimes joined up with the Slough pack

Light Gray and Dark Gray, two young males who were interested in mating with the young Druid females

Other Packs Mentioned

Blacktail pack

Geode Creek pack

Lava Creek pack

Leopold pack

Oxbow pack

Quadrant pack

Silver pack

Unknown pack

PROLOGUE

T HE OLD WOLF ran all out. If he could just get a little farther, he would be safe. But he sensed that his pursuers were gaining on him. In a few moments they would reach him. He tried to speed up, but his lungs were reaching their limit and he gasped for breath. He had to make a decision. Should he keep on running for however many more seconds he could, or should he turn around, face what was coming at him, and fight it out?

PREVIOUSLY
IN LAMAR VALLEY

———

P ARK RANGERS KILLED the last of the original wolves
in Yellowstone in 1926. By 1995 attitudes had changed.
In 1995 and 1996, thirty-one wolves in seven different
packs were brought from Canada and released to reintroduce
a native species to the park. After a few years, Yellowstone
averaged about one hundred wolves in ten packs distributed
over 2.2 million acres.

I started working in Yellowstone in the spring of 1994 and
was designated the park's Wolf Interpreter. At the time I was
the only ranger who had been around wild wolves in other
parks. For fifteen summers I had worked in Denali National
Park in Alaska and had spent another three summers in Gla-
cier National Park in northern Montana. All my programs
that first year in Yellowstone were about wolves and Yellowstone's Wolf Reintroduction Program.

After the first wolves were released in March of 1995, I
continued to do wolf talks and spent much of my time
watching the wolves and studying their behavior. We had
not expected that they would be very visible to the public,
but one of the new packs was often in sight from the park

road. When I was on duty, I would find the wolves and help people see them through my spotting scope. I did the same thing on my days off.

In the spring of 1998, I switched from public outreach with the Naturalist Division to working for Yellowstone's Wolf Project, joining the research team that studied the thriving wolf population. The Wolf Project radio-collars a number of wolves every year, with the goal of having at least two wolves with functioning collars in each pack. The project's staff studies the wolf population and puts out research papers along with annual reports summarizing what we have learned about the packs and individual wolves. Doug Smith, the program's senior scientist, was my boss.

My new position also involved working with the public, so I continued to help people see the wolves and gave frequent talks along the park road as we watched the packs. I did most of my wolf observations in Lamar Valley, along the road that leads from Silver Gate, Montana, at the Northeast Entrance to the park, to Mammoth Hot Springs, south of Gardiner, Montana. The eastern end of that section of the park was the area where we most often saw wolves.

The Druid Peak pack was the most prominent wolf family in the early years of the reintroduction, and the pack grew to thirty-eight wolves at its highest point in 2001. The alpha pair, wolves 21 and 42, became famous worldwide because of the high visibility of their pack in Lamar Valley and television documentaries on their lives. 21 was the epitome of what an alpha male wolf should be and diligently took on the responsibilities of leading hunts, defending the family from rival packs, and helping to raise pups. I learned that the alpha female is the true leader of the pack, and 42 proved to be highly competent in making major decisions such as where to den and when to move the family's pups to other locations. As I watched 21 and 42, I could see that they were very devoted and bonded to each other.

In early 2004, alpha 42 died. 21 was an old wolf by then and was never the same after the death of his longtime mate. He took off by himself a few months later. We found his remains in a high meadow above Lamar Valley, a place he and 42 often visited together. 21's apparent successor was his adult son, wolf 253, who had been well trained by his father. But 253 had two injured legs, which would be a big liability if another wolf challenged him for the Druid alpha male position.

The Agate Creek pack lived to the southwest of the Druids. It had been founded in 2002 by several of 21's daughters and male 113 from a pack in the northwest section of the park. The Slough Creek pack, started by another of 21's daughters, lived west of Lamar Valley.

By the time this story starts, the Druids were no longer the superpack they had once been. With the passing of 21 and 42, those of us watching the wolves felt the dynamics between the three neighboring packs could change drastically. It seemed likely that the Druids' next alpha male, whether it was 253 or a new wolf, would have to contend with attempts by rival wolves to annex parts of the family's high-quality territory in Lamar Valley.

PART 1

2004

1

To Be the Alpha
You Have to Beat
the Alpha

AFTER LONGTIME DRUID alpha male 21 died, two black males showed up in the pack's territory. One was a wolf I knew well: 302. He had been born into the Leopold pack, and the Druids' new top male, 21's son 253, was his cousin. Both males were four years old. 302's companion, a yearling Leopold male, was 302's nephew.

The wolves in the Druid pack were already acquainted with 302 because he had made frequent visits to the Druids the previous year and gotten several of 21's daughters pregnant. Their father took an instant dislike to 302 and tried to chase him off. I had seen 21 catch 302 and beat him up but then let him go. That reminded me of how 21 was raised by a stepfather, wolf 8, who never killed a defeated opponent. That

3

role modeling apparently had deeply influenced the young 21 for he grew up to be a fierce defender of his own family, yet always spared the lives of defeated interlopers like 302.

302 soon left those females and returned to his own pack. 21 and 253 ended up raising five pups sired by 302, along with many pups fathered by 21. Wolf 302 must have had some sense of paternal duty, for he regularly came back to Lamar Valley to visit his females and their young pups, then went back to the Leopold territory, twenty-five miles to the west.

After the death of Druid alpha female 42, one of her daughters took over that top position. Two of the Druid females, including the new alpha, had pups in the spring of 2004, presumably fathered by 21 before he left the pack to die alone in that high meadow. The new alpha female had her pups north of the park road at the pack's traditional den site. The other mother denned in an area south of the road known as the Chalcedony Creek rendezvous site. The Chalcedony site was where the Druids traditionally moved their pups once they were old enough to travel. It had water and a meadow where the pups could play together and learn to hunt small rodents and insects. The adults would usually leave the pups there with a babysitter or two while they went off to hunt.

253 spent most of his time at the alpha female's den, which gave the two Leopold wolves opportunities to visit the other young mother at the Chalcedony site without worrying about running into the Druids' alpha male. With two big adult males in the valley, the big questions were: Would 302 challenge his cousin for the Druid alpha male position? And if he did, who would prevail?

Early on June 14, I picked up the signal from 253's radio collar in the direction of the main Druid den. The two Leopold males and other Druids were in the rendezvous site across the road. I heard from other wolf watchers that 253 had crossed the road to the south, then I spotted him approaching the rendezvous site. The two Leopold males saw 253 and immediately charged at him with their tails raised. The three males met up in tall sagebrush and fought. This was it: the moment 253 would have to prove he was worthy to be the Druid alpha male.

The sagebrush partially blocked my view but based on what I could see and what I later saw in a video of the fight shot by a local man named Cliff Brown, the key moment in the battle came when 253 and 302 were in a standoff. At that moment, 302's young nephew ran in and the two Leopold males stood on either side of 253. The Druid male quickly made the first move. He lunged at 302 and bit him. 302 ran off with his tail tucked. Then 253 turned and went after the young nephew, who also ran away.

Other Druids joined 253 and they chased the younger outsider. The Druids caught up with the Leopold wolf, pinned him to the ground, and bit him. 302 ran back toward his nephew and did something that astounded me. He joined the Druids in biting his nephew while he was down. The yearling wriggled out from under his attackers and ran off. The Druids pursued him, with 302 running among them. If 302 thought that would get him on the good side of the Druid alpha male, it did not work. 253 turned around and charged at 302.

The three males soon were in another standoff. This time, 253 lunged at the yearling Leopold as 302 charged at 253

from behind and bit him. The Druid alpha male fought back
and 302 ran off in fear. At that moment, the younger Leopold ran at 253. 253 spun around and went after him. He
caught the wolf and, despite his injured legs, easily defeated
and pinned him. Other Druids ran in and all of them nipped
at the Leopold yearling.

302 was thirty yards away, watching the other wolves
attack his packmate. After some hesitation he raced over and
once again joined the Druids in biting at his nephew. Then he
ran off. The Druid alpha male bit into the yearling and shook
his head from side to side. 302 ran back and once again bit
his nephew. I had never seen such bizarre behavior in a wolf.

At that point, the yearling managed to jump up and run
off. The Druids chased him, but he outran the pack. He saw
302 and, despite how he had just been treated, ran to him.
The two Leopold males sped off to the west, then paused to
look back. The Druids were still charging toward them. The
interlopers resumed their flight. The Druids soon stopped
and watched as the two males left the area.

The Druids went back to the rendezvous site. Their new
alpha male had defeated two opponents that morning, both
big males. Through personal combat, he had earned the right
to lead the pack. But he was limping badly on his front right
paw. He held it off the ground as he walked around the scene
of the battle. That paw had been caught in a steel trap when
he went out of the park in late 2002, over nineteen months
previously. He was probably bitten there during the battle
and the bite caused his old wound to flare up.

Thinking about how well 253 fought that day, I recalled
how he had apprenticed under 21 who was undefeated in

combat. When father and son engaged in roughhouse play and wrestling, 21 would have demonstrated methods of defeating a rival wolf to 253. I recalled seeing 21 engage in matches with his sons and letting them take him down to the ground. It would be like Bruce Lee doing something similar with his sons. 253 had learned from a master of martial arts and the training saved him that day.

In the days after the fight I thought a lot about 302 and what I knew about him. 302 had a history of getting females pregnant and abandoning them. He seemed to be afraid of getting into fights with other wolves, even when his females were threatened, and he usually ran away so he could save himself. I thought of him as being like men who have Peter Pan syndrome, a condition where they never seem to grow up or take on adult responsibilities.

There was, however, a better way to think of 302: he was a renegade. He seemed to be in rebellion against the traditional roles adult males play in wolf society. As I have seen so many times over the years, the normal path in life for a young male wolf is to seek out a mate, find a vacant territory, then devote himself to raising pups with his mate year after year. Two critical aspects of being a traditional alpha male wolf are the willingness to risk your life to support your family by hunting prey animals that are usually much bigger and stronger than you are and a commitment to fight any rival wolves that threaten your mate and pups. 302 was especially lacking in that department for he usually gave his priority to saving himself, rather than protecting females he had gotten pregnant. I once had the thought that the Bon Jovi song "You Give Love a Bad Name" could have been about 302.

The book that most influenced me to appreciate nature and wildlife was Henry David Thoreau's *Walden*. He wrote that some people step to the beat of a different drummer than everyone else. That was true of Thoreau himself who did not conform to the expected role of men in society at that time. 302 was a free spirit who marched to a beat that was different from the other male wolves I had known.

2

The New World Order

21'S SON HAD won the right to be the Druid alpha male in combat, but as the summer progressed he seemed to have only a tenuous association with the pack. From July 15 to August 31, he was in Lamar Valley twice, for just one day each time. I thought about 253's relatedness to the various Druid females. They would be his full sisters or half sisters. It was unlikely he would be interested in any of them during the February breeding season, now little more than six months away, as wolves seem to have an instinct to avoid breeding with close relatives.

I think 253 understood that and was trying to find an unrelated female outside Druid territory. If he found one, there would be a problem if he brought her home. The Druid females would be aggressive to her. If 253 could not get them to accept his new mate, the pair would have to seek out a vacant territory and start a new pack. 253's departure would clear the way for 302 to take over as the new alpha male. Even with his nephew's help, 302 had failed to defeat 253, who had proved to be too formidable a fighter despite his injuries—but 302 might get the Druid alpha position by default if he could just stick around long enough.

On August 1, I spotted 302 with the Druids in the rendezvous site. As the pack traveled east, he and the Druid alpha female did six double scent marks in an hour, a sign that she regarded 302 as the pack's new alpha male. 302's nephew and the young Druid males also treated 302 deferentially. That same day, however, I saw 302 sniff a grass clump 253 had scent marked three days earlier. He circled around the spot, acting like he was afraid to lift his leg at the site to assert an alpha male status. It appeared that 302 was not totally secure in his new position and worried that 253 could come back at any moment.

The Druid alpha female, wolf 286, was often affectionate and playful with both 302 and his nephew, who became known as wolf 480, but the younger male was more responsive with her than 302. One day she approached them both in a playful posture and chased 480. She nipped him on the rear end during the pursuit. They were both acting like young carefree pups. The much older 302 did not join in.

It was still unclear whether 302 felt he really was the Druids' new alpha male. When 253 turned up at the rendezvous site on August 26, both 302 and 480 stayed away. The two males ventured back three days later and found the site empty. They did a lot of howling, trying to contact the pack, but got no responses. 253 had led the Druids up the Lamar River and there was no sign of them until September 2 when we saw 255, one of the older females, traveling with 302 and 480 to an area west of Slough Creek. This area of lush meadows, hills, and ponds was known as Little America, and wolves of several neighboring packs often visited the site. For that reason it became a popular spot with tourists, who would pull over and watch the wolves from the roadside.

Four days later, 255 was back at the rendezvous site, where 286 had brought her pups to join the other litter already based there. We had previously spotted seven pups at the Druids' main den and two at the den at their rendezvous site, but 255 was babysitting just four pups, which meant that the other five pups from the Druids' two litters had likely died.

255 was overseeing the pups while the rest of the Druids were elsewhere, probably on a hunt. She was the oldest female in the family so was acting like a human grandmother taking care of the family's young children. Her presence might be critical to the remaining pups' survival for grizzlies often came through that area.

That is exactly what soon happened. A grizzly sow with two first-year cubs walked into the rendezvous site. The pups raised their tails excitedly and charged at the bear family. 255 saw what was happening and joined the pursuit but the grizzlies were already on the run from the pups. Bears have poor eyesight and the mother grizzly must have mistaken the pups for adult wolves. After that diversion, the pups chased each other and played tug-of-war with a strip of fur. When one pup got tired of communal play, it caught and ate grasshoppers.

Two days later, 302 and 480 arrived with alpha female 286. She passed the time there by chasing small mouselike rodents called voles. 286 would catch one, let it go, and pursue it again. 480 tried to figure out what she was up to. He walked over, lay down, and watched the alpha female acting like a pup. She caught another vole, shook it, then tossed it fifteen feet through the air. It ran off as soon as it landed. 286 grabbed the vole and flung it around ten more times. After the last toss she tried to get it to flee by pushing it with her nose, but the last crash landing had killed it. She carried the little animal toward 480 and dropped it close to him, like she was daring the big male to take it from her. 480 ignored the vole, but he did get up and go over to her. She teasingly romped around him. 302 came over, looked over the area, and ate her vole before she could run back. He had a talent for quickly figuring out how to take advantage of a situation.

The Wolf Project did regular tracking flights, usually once a week or so, to monitor collared wolves and their packs. When a pack was spotted, the plane would circle the wolves, and the Wolf Project biologist on board—usually Doug Smith or Dan Stahler—would count the adults and pups, note any kills or other information on what the pack was doing, and record their location. As the biologists needed to cover most of the 2.2-million-acre park, those flights often lasted four hours. There was no bathroom on board, so they had to plan ahead. Our pilot was Roger Stradley, who eventually accumulated over fifty years of flying experience in Yellowstone. Unlike some pilots, he had a smooth style of operating the plane when circling and flying in windy weather that minimized the chances of passengers getting sick. For that and many other reasons, he was much loved by our staff.

On the September 21 flight, Dan did not get signals from 253 in the park. Another flight a week later also failed to find him. I thought about the time 253 left the park a few years earlier, went to Utah, and stepped in a coyote trap. He was turned over to the U.S. Fish and Wildlife Service. His injured paw was treated, and he was released near the South Entrance of Yellowstone. 253 walked back to Lamar Valley and rejoined the Druids. There were tracks of a second wolf in the area where he was trapped and it was likely a female he had just paired off with. Perhaps 253 was heading back to Utah to look for her.

302 was destined to have a similar encounter with a leg-hold trap. A local wildlife organization was conducting a coyote research project. Their staff had permission to set out traps in Lamar Valley and other locations so coyotes could be caught and radio-collared. I got a report that a black wolf had stepped in one of those traps by the Lamar River.

I drove to where I could see the site and identified the wolf as 302. His left front paw was in the trap. He was digging at the cable anchoring the trap with his other front paw, frantically trying to free himself. I noticed there were a lot of bison coming into the area. At risk to themselves, two of the coyote researchers twice diverted groups of bison that were heading toward 302. If they had reached him, the bison would have gored the trapped wolf to death or trampled him. Likely all of those bison had been chased by wolves and some of them might have seen a herd member killed by wolves. They would gladly kill a wolf that appeared to be defying them by standing its ground.

I got word to Doug and Dan and they soon arrived from park headquarters. We hiked out to the river, waded across, then continued toward 302. Doug was carrying a big fishing net. He planned to capture the wolf in it, then tranquilize him. As Doug got closer, 302 jumped up and struggled to pull his paw out of the trap. Just as Doug was about to net him, 302 yanked his paw free and ran off. He limped on that paw for many days, then it seemed to heal. The Park Service revised the coyote trapping permit after that incident and required that traps not be placed on wolf travel routes. 302 had stepped in one on just such a route.

In early October, several of us who had known and admired former Druid alpha female 42, alpha 21's longtime mate, hiked up on Specimen Ridge to the site where she had died the previous winter. We found only one rib at the location. A Wolf Project crew had retrieved her skull soon after her death for research purposes. Scavengers had carried off the rest of her remains.

We shared stories and memories of 42. I thought about how both 42 and 8, the wolf who had adopted and raised 21, were picked on and bullied by their siblings. When they later became alphas, 8 and 42 used cooperation and teamwork as their primary leadership methods rather than aggression. 21 spent two years being mentored by 8, then six and a half years as 42's mate. 8, 42, and 21's mother, wolf 9, were the three most important influences on making 21 such a great alpha male.

How could any Druid alpha male compare to the legendary 21? 253 carried his father's genes and had been trained by him. He could have been a worthy successor but had left the family to find a mate. I doubted that 302 could ever measure up to what most people consider to be Yellowstone's greatest wolf.

3

480's Trial
by Combat

ON JUNE 30, Doug Smith did a monitoring flight. He got a mortality signal from alpha male 261 of the Slough Creek pack, the group based just west of the Druids. The wolf's body was found next to a creek. Too much time had passed to determine the cause of death, but his presence by water suggested that he had been injured, probably by an elk that had waded into the water to escape 261, and the wolf had stayed by the creek as he tried to recover. One of his brothers, wolf 490, became the next Slough alpha male.

The Slough Creek pack had been formed at the very end of 2002 by 217, one of 21's daughters, and 261, who had been born into the Mollie's pack. The Mollie's wolves were descendants of the Crystal Creek wolves, one of the first three packs to arrive from Alberta in 1995. They had been driven out of Lamar by the formidable Druids a few months after that pack was brought in from British Columbia in 1996 and eventually established a new territory to the south in Pelican Valley. The snow was so deep in that part of Yellowstone that elk normally left the valley in the winter. But the local bison stayed put, meaning they were the only prey available to the Mollie's wolves. Male members of the pack tended to be larger than other wolves in the park, most of which focused on hunting elk, which are much smaller prey than a 2,000-pound bison. Having several big males in the pack gave the Mollie's wolves an advantage when hunting bison and defending the pack from rival wolves.

With the death of 261, another big Mollie's male in the pack stepped up to lead the Sloughs. The Slough founding alpha female, 217, had died in January and an uncollared black was now the top female. She was likely yet another of 21's daughters. The pack usually had seven to eight adult members, about the same number as the Druids.

In late September, I saw the Slough wolves on a hunt near their namesake creek. Six wolves raced along the creek bank, paralleling a cow elk who was running through the water. Several wolves jumped in the creek and went after her. But the water was deep there, and the wolves had to swim while the much taller cow could still run through the creek. Wolves, like many dogs, are good swimmers, but it is more difficult for them to attack their prey when they are swimming rather than running. Elk seem to know that and usually try to escape wolves by plunging into the creeks and rivers of Yellowstone.

The cow reached a shallower area and the pursuing wolves caught up with her. Two of them got in front of her. The cow leaped over them, then reached a deeper section of water. A big gray male swam after her and grabbed a hind leg. She violently bucked up and down, trying to shake him off. A black wolf swam over and got up on her back. It bit at the back of her neck, where an African lion would make the killing bite. Another black attacked her hindquarters. More wolves swam out. The cow collapsed in the water as the wolves fought with her. The wolf on her back was still biting her neck. It had placed a front paw around her throat to help maintain its balance. Another wolf bit at the side of her neck. The cow tipped over and floated on her side, but still kicked out with her hind legs. A minute or so later all movement stopped.

Veteran wolf watchers Mark and Carol Rickman stayed with the pack after I left and later told me wolves would leave the carcass after stuffing themselves, trot off to the north, come back to the site and feed, then go north once more. They must have been bringing food to their pups. A few days later, I saw eight pups with the Slough adults.

In the past, the Slough wolves had typically stayed out of the Druids' way, but I wondered if the relationship between the two neighboring packs would become more competitive with the death of alpha female 217, the former Druid, especially as the Sloughs had so many surviving pups this year in comparison with the Druids.

In mid-October, there was a skirmish between the Druids and three Slough Creek wolves, including the Slough alpha pair. The Druid group included 302, his nephew 480, alpha female 286, and five others. Several of them pinned a Slough wolf and bit at it, then let it run off. All three Slough wolves raced across the road and the Druids did not pursue them. Those Sloughs later reunited with the rest of their pack. That incident took place near the border of the packs' territories. If the Slough wolves continued to intrude into Lamar Valley on hunts, there would likely be more serious confrontations. The big question now was how 302 would function in a battle with a larger group of rival wolves. Based on his track record, I had my doubts that he would be capable of successfully standing up to an aggressive invading force.

On the morning of October 26, I found eight Druid adults and two pups bedded down in their Chalcedony rendezvous site. I got signals from two big males from the Mollie's pack, 378 and 379, to the west. Those brothers often came north to visit their relatives in the Slough Creek pack. A short time later, the two Mollie's wolves ran at the Druids with their tails raised. They singled out a sleeping wolf and violently attacked him. It was 302. As the other Druids chased off 378, wolf 379 continued to attack 302, who now was fighting back as best as he could. Then 302 gave up and ran away to the east.

A group of Druids chased 379 in the other direction as 302 kept running, not the proper behavior for a pack's alpha male. The primary duty of an alpha male is to protect his family from rival wolves, and 302 was failing miserably in that task. In contrast, his young nephew was one of several Druids chasing the two males away. I later saw 302 across the river, alone, still distancing himself from his pack. It was a sad sight for he had let his new family down when they most needed him. That was especially significant since DNA testing showed that he had sired five of the pack's yearlings. He had not just taken over an unrelated pack as the alpha male. Many of these wolves were his sons and daughters.

The next morning one of the Mollie's brothers was still in the area, and 480 challenged him to a fight. 480 was wounded in the battle but courageously kept fighting back until he won and chased the much bigger 379 out of the area, getting revenge for the previous day's attack on 302. That was the ultimate test for an alpha male wolf: not giving up when things got tough.

It was becoming clear that 480 had the right stuff to be an alpha and 302 did not. That position and its responsibility would be a heavy burden for 480. I once heard that in ancient China it was considered a curse to say to someone, "May you live in interesting times." During China's long history, interesting times meant invasions by enemy forces, tyrannical rulers, forced exile, floods, and epidemics. Better to live in boring times. 480 was destined to be an alpha male living in interesting times.

Doug did a flight a few days later and saw that 378 and 379 had gone back to the Mollie's pack territory. I figured they had realized 480 was too much for them so had given up on trying to take over the Druid pack.

302 returned to the Druids two days after he had run off, holding his hind right leg off the ground and limping on a front leg. He must have been bitten on both legs by the Mollie's brothers. When 480 came over, he held his tail higher than 302's, a sign that he now considered himself the dominant wolf. The Druids soon moved off and 302 struggled to keep up. He often bedded down and licked a hind paw, then a front paw. When he got up and followed the others, he hopped on his good hind leg so he could keep the injured one off the ground.

A few days after that, 302 went to 480 and licked his face as he wagged his tail from a crouched position, acknowledging his young nephew as the Druids' new alpha male. When the two Leopold males had arrived in Lamar Valley that spring, 480 had acted like a sidekick to 302, like Batman and Robin. Now their roles were reversed: 480 was the heroic main character and 302 had been relegated to a supporting role.

A few days after 480 proved himself in battle with the Mollie's wolves, I saw another side to the Druids' new alpha male. A black pup romped over to him and bit the fur on his shoulder. 480 tried to walk away so he could rest. The pup continued to harass him and the big male playfully ran off, but at a slow pace. The pup chased after him, then grabbed 480's tail and tugged at it. 480 continued forward, with the pup attached to him, then suddenly stopped, causing the pup to trip and fall. 480 never did anything to discipline the pup. I thought of the contrast between that incident and the all-out fight when 480 beat 379. When needed, he could be the toughest wolf around, and at other times, he could be extremely tolerant, even with a pup he had not sired. All that suggested that 480 was going to be an alpha male in the tradition of the great 21.

In mid-November, we saw eleven Druids running west along the top of Specimen Ridge above Little America. They were chasing nine members of the Agate pack, their neighbors to the west. Wolves ran back and forth in confusing patterns. Soon the Druids came together for a rally.

The Agates stopped and their alpha male, 113, howled at the Druids. This part of Specimen Ridge was his territory and the Druids were invading it. Soon the Druids turned around and headed back toward Lamar. Some of the younger Agate wolves ran after them, which set off more chasing between the two packs. The Agate alpha female seemed to hang back. She had grown up in the Druid pack so may have hesitated to go after her relatives.

There was more running back and forth, but the Druids, with more wolves in their group, did most of the chasing. The skirmishes continued for over two hours. The scene might have been like watching a Civil War battle, where one army chased the opposing soldiers, then that force regrouped, turned around, charged forward, and pursued the army that had just chased them.

The Agate alpha female finally had enough. She took charge and led west, away from the Druids, and the rest of the Agates followed her lead. I realized there was no actual fighting during the lengthy incident. It was all just chasing. I was impressed with that female's actions. If the skirmish had continued, some wolves might eventually have been injured, and with the Agate pack outnumbered, it would likely have been one of their members. The episode reminded me of some news coverage I had seen of a riot in a big city. A mother burst on the scene, found her teenage son, who was a foot taller than she was, slapped him in the face, and dragged him away. He meekly went along with her. This incident between the Druids and the Agates was one more case that showed that mother wolves are the real bosses in the family.

I also thought about 480's behavior during the interaction. He chased Agate wolves but seemed to run at well under his top speed. Since I had witnessed his fight with 379, I knew he was very capable in combat, but in this situation he never touched an Agate wolf. The more I watched him, the more I felt he had a personality that was much like 21 in that neither big male resorted to violence when it was not needed.

A few days later, I got a report of another skirmish from Tim Hudson, who was on one of the Wolf Project's Winter Study teams. Each winter three wolf packs are monitored intensely for two thirty-day periods, one in early winter and another in late winter. Tim was watching his study pack, the Geode Creek pack. Nine of them had killed an elk at Hellroaring Creek and were feeding on it when ten Leopold wolves charged in from the southwest while another ten Leopolds approached from the north. It looked like a World War II pincer operation. The outnumbered and outflanked Geode pack split up and ran off in different directions, an effective counterstrategy. I joined Tim's crew and counted twenty-two Leopolds around the elk carcass. Later those wolves followed the scent trail of the Geode wolves and chased some of them but, similar to the encounter between the Agates and the Druids, no wolves were harmed.

That same day, the Druids had another encounter with a neighboring pack. My co-worker Emily Almberg and her friend Mike O'Connell witnessed a battle between the Druids and part of the Slough Creek pack. At one point in the lengthy conflict, 480 and five other Druids caught and pinned a Slough pup and nipped at it. Two Slough adults separately tried to rescue the pup, but the Druids drove them off. Soon after that, the six Druids left. Emily saw the pup get up and walk away—it appeared that the Druids had not seriously harmed it. During the reign of 21, we had never known him to kill a defeated rival. So far 480, the new alpha male, was continuing that tradition.

Not all alpha males are so benign, which meant that the two Slough adults who tried to rescue that pup were risking their lives. The Wolf Project documented a number of similar cases, and Kira Cassidy, a Wolf Project biologist, and I wrote a peer-reviewed research paper on altruistic behavior we had witnessed in Yellowstone wolves, mainly rescues of fellow pack members who were being attacked by rival packs.

In late December, the Slough wolves were frequently spotted in Lamar Valley. One day I saw fifteen of them at the Druids' Chalcedony rendezvous site. Seven Druids were watching them from the west. A black Druid broke off and fled farther away from the Sloughs. It was 302. The other six Druids, including 480, ran toward the rival pack. The Sloughs charged and the Druids, realizing they were heavily outnumbered, scattered. The Druids' strategy of splitting up and disappearing into the countryside served them well. The Slough wolves soon lost interest and traveled west toward their territory. Another territorial skirmish had passed without any fatalities on either side.

I later spotted the Druids up high on Mount Norris, the peak south of their traditional den site. From there they had a good view of any approaching wolves. That day the pack acted like a guerrilla band who used their intimate knowledge of their homeland to slip away from a superior invading army.

At the end of 2004, the biggest wolf pack in Yellowstone was Leopold with twenty-three members. The Sloughs numbered fifteen, and the Agates eleven. The Druids had seven adults and two surviving pups, both females—a black and a gray. In three years, the pack had gone from being the largest ever seen in observations of wild wolves, with a count of thirty-eight, to the smallest one in their section of the park. The pack would soon get even smaller.

As the year was ending, I thought about 253. He had last been spotted at Tower Junction on October 24. We found out later that he left Yellowstone shortly after that and traveled south, toward Jackson, Wyoming. It would take more than a year, but eventually I was to get news of how 253 was faring after he decided to leave the Druids for good.

I also took a few moments to think about the Druids. The pack was now led by 480 and 286 with 302, the former alpha, reduced to being an auxiliary male, subordinate to his much younger nephew. As the next mating season approached, would he be willing to accept that demotion and stay in the family, or would the sound of a distant drummer cause him to take off and try to find a better situation?

PART 2

——◆——

2005

4

Outnumbered

AS I WATCHED 302 in early December, I saw that not only was he subordinate to 480, he was now lower ranking than a black yearling male. I saw him lick that black's face while the yearling held his tail higher than 302 held his. Out of four adult males, he had fallen to the third-ranking position, even though he was the oldest wolf in the pack. Later in the month he and that yearling fought twice and 302 lost each time. To put it kindly, fighting was not 302's strong suit.

Despite his downgraded status in the pack, 302 was likely hopeful of breeding in the approaching mating season. Normally an alpha male guards the alpha female and other unrelated females from lower-ranking males that might try to mate with them. But 302 was popular with females and had a successful track record of getting away with things. I suspected he was going to have a successful mating season despite his reduced status in the pack.

On January 5, wolf 302 sniffed a young Druid female, then tried to mount her, but she turned and snapped at him. It was far too early for a mating. After that he went to alpha female 286 and twice mounted her, but each time she spun around and threatened to bite him. 480 was nearby but did not intervene, perhaps because he saw the female's aggressive behavior to her suitor. 302 tried three more times to mount 286 and she rejected him during each attempt. 255, an old mating partner of 302's, came over and was more friendly to him, as was female yearling 375.

On January 12, 2005, we had a tenth anniversary celebration of the arrival of the first batch of wolves for the reintroduction project. I thought about how much we had learned about wolf behavior in the past decade and the many life stories of extraordinary wolves we had come to know. I also reflected on how wolf packs, like human empires, can increase in size and expand their territories but later experience a significant decrease in members when times change. The Druids had been the largest wolf pack ever known but now were only average in size. I expected one of their neighboring packs would take advantage of their diminished numbers.

A few days after that celebration, the large Slough Creek pack traveled through the north side of Lamar Valley. Later seven Druids howled from south of the road. Almost instantly we heard wolves howling from the Druids' traditional den forest north of the Footbridge parking lot; signals indicated the sound was coming from the Sloughs. Having a rival pack confidently howling from their main den site must have greatly disturbed the Druids. Later I saw the Sloughs coming out of the den forest. They howled again as they looked in the direction of the Druids.

The Slough wolves apparently thought they were so numerically superior to the Druids that they could freely walk into their den area and howl repeatedly to announce their occupation of the site. The next day, the Sloughs were still there and I failed to get any Druid signals in the valley. When we had lost track of the Druids in the past, they were usually up the Lamar River. I suspected that was where they had retreated in response to the Slough pack's bold incursion into the heart of their territory.

That afternoon I saw 378 and 379, the two brothers from the Mollie's pack, south of the Druid den forest. They were the males who had attacked 302, and 480 later defeated 379 in a drawn-out fight. For now, those two wolves were part-time members of the Slough pack. If they joined the pack permanently, the Sloughs would number seventeen.

Most of the Druid yearlings had left to look for mates, and the pack was now down to just seven: five adults and two pups. That count did not include 376, the mother of the pups born in the Chalcedony rendezvous site. It had been a while since we had seen her, and because the battery in her collar had died, we could not track her. We never knew what happened to her, but I suspect that she had run into the Slough wolves and been killed. The Druids were at a low point in the family's history, far removed from their glory days when they were the dominant pack in the region.

I watched as the black Slough alpha male easily pinned his brother 379. If he could do that, he could probably beat 480 in a fight. There were two other big Mollie's males among the Sloughs. Since 480 could not count on 302 backing him up if he had to defend the pack, he might have to fight four big males at once. Would he run off if he was confronted by those four males? Or would he stand his ground and be killed in a fight where the odds were overwhelmingly against him? I guessed he would not run away.

The Sloughs stayed near the main Druid den site for the next two days. By the morning of the third day, they had moved south of the road. The Druids soon returned to that forest, where the family had denned since 1997 and where eight generations of Druid pups had been raised. I imagined them sniffing around all the places the Sloughs had been. I had no doubt that the alpha female would be very concerned if she had planned to den there. Would she still use the site or choose a safer place to have her pups that spring, a site farther away from the Slough wolves?

Later that January, the Slough wolves once again did a lot of howling in Lamar Valley. The Druids defiantly howled back but wisely stayed away from the rival pack. The Sloughs remained in Lamar for six days. It looked like they intended to annex that part of the Druid territory. The Druids would have to keep to the far reaches of their homeland or abandon what had once been theirs and move elsewhere. The increasingly bold Sloughs were the biggest threat the pack had ever faced.

5

Hard Times
in Lamar Valley

I N JANUARY THE Druids were on a carcass south of Slough Creek. Fifteen Slough wolves charged at the seven Druids. After driving the Druids off, the Sloughs took over the kill site and fed. Cinematographer Bob Landis filmed the incident, and I later studied his footage. I was not surprised to see that 302 was the first wolf to run from the Sloughs. 302's behavior in conflicts made 480's situation even more challenging. Not only was his pack outnumbered by the Slough wolves, he could not count on 302, the biggest male in the Druids, to back him up if the two packs got into a fight.

By that time, Doug Smith had darted and collared four of the Slough wolves, including alpha male 490, the beta or second-ranking male 377, and the third-ranking male 489. The alpha female remained uncollared.

In smaller wolf packs, the alpha male is usually the only male to mate, either with the alpha female or with her and other receptive lower-ranking females in the group. But in large packs where there are several adult males and adult females, things can get complicated. The mating season for the Sloughs started in late January when third-ranking male 489 bred an uncollared subordinate gray female. The alpha pair saw them get in a breeding tie, ran over, and pinned both wolves, but they were too late to stop the mating. Seven minutes later, the alpha pair got into a tie. A few days later the beta male, 377, mated with a black female, and part-time pack member 379 got in a tie with another black. In both cases, the alpha male saw the ties only after they were in progress, too late to intervene. On another day, 377 got away with breeding the alpha female.

The Wolf Project staff documented other cases where lower-ranking males were successful because they cleverly took advantage of opportune moments. In the Canyon pack, the beta male ended up on one side of a fence in the Mammoth area while the alpha male was on the other side. The beta went to the nearby alpha female and quickly got into a tie with her. The alpha male frantically tried to jump over the fence or get around it but failed to reach the pair in time to break up the mating. Later the alpha male was darted in a collaring operation and the other male took advantage of his temporary incapacity to mate with the female again.

I expected to see some similar situations in the Druid pack for 302 was an expert at picking the right moment to make his move. There were three females of breeding age in the Druid pack—alpha 286, the older female 255, and yearling 375—and all apparently wanted to get pregnant. It was a full-time job for 480 to keep 302 from them. One day in early February, 286 was leading the pack toward the road, intending to cross to the north. A truck pulled up at her crossing spot and 286 stopped to evaluate the situation. 480, who tended to be the wariest wolf in the pack, was well behind her.

302 had little fear of the road and vehicles. He paused, looked at 286 and the truck, then ran forward, jumped on 286, and quickly got in a breeding tie. 480 saw what was happening and raced to them, but he was too late. 286 snapped at the alpha male, a message not to interrupt her. I reviewed Bob's footage of the incident and saw that 302 was in a tie with the female within eleven seconds of reaching her. He could work fast when necessary. Sometimes life is unfair. 480 was a classic alpha male worthy of taking 21's place in the Druid pack, but the Druid females seemed to be more attracted to 302 than they were to him.

The relationship between the two related males reminded me of a letter I read in a Dear Abby column. A man wrote in about his brother Chaz who "has always been extremely popular with everyone. He is the better-looking, more talented, smooth-talking brother. Ever since I can remember, people have walked past me on their way to flock around him. The only time girls talk to me is when they ask about him. It hasn't been easy living in his shadow and being seen by everyone as 'just his brother.' It has done a real number on my self-esteem." 302 had the same charisma as Chaz.

While all that mating activity was going on, the dispute between the Druids and the Slough wolves continued, and the Sloughs were winning. On February 10, Doug did a tracking flight. He called down to say he was getting a mortality signal from Druid female yearling 375 up the Lamar River.

We tried to figure out what had happened. Druid and Slough wolves were in that general area at that time, and during the flight, Doug saw some of the Sloughs on a bull elk carcass. He and other Wolf Project staff later visited the site. It looked like the Slough wolves had spotted the Druids feeding on the elk and charged in. Tracks showed 375 fled upriver and was attacked at several spots. She apparently got away each time for Doug found her bloody tracks in the snow. She likely later bled to death.

I checked my records and pulled together some information. On February 9, I had seen the Sloughs going east through Lamar Valley. Later in the day, signals indicated they were traveling up the Lamar River, where the Druids had retreated earlier when the Sloughs occupied their den site. We had gotten the Druid signals that way, so it looked like the Sloughs were tracking them down. The next morning Doug got the mortality signal from 375 in that area.

The incursions of the Sloughs into Druid territory continued. Early on the eleventh, I spotted fourteen Slough wolves south of the Druid den forest. Late that morning we found four Druids five miles to the west: 480, 302, the older female 255, and the black pup. The alpha female, 286, and the gray female pup were missing.

The next morning, both missing wolves turned up at the Chalcedony rendezvous site, but they were not alone. Two big males from the Slough pack were with them: 379, the Mollie's male who had fought with 480, and 489.

I had never seen anything like this situation before. The Druid and Slough wolves were rivals and the Sloughs had just killed a Druid female, but here was 286 with two of their males and she seemed to be acting friendly with them. I had to conclude that they were interested in mating with her, which apparently overrode the past history between the two packs.

Later 286 and the pup went west, toward the main Druid group. 379 followed them. I had the impression that the two Druids wanted to get away from him. The pup twice lunged and snapped at the big male when he approached, but he tolerated her actions. When 379 went to 286 she responded very aggressively. He followed them for several miles, then ran off when the two Druids headed toward the road. By that time, the other four Druids were at Slough Creek. I went there and saw 480 breed the older female, 255. Later 286 and the gray pup joined them.

The Druids spent the next few days up the Lamar River, likely looking for 375. Probably all they knew was that she had been missing since the confrontation with the Slough wolves at the elk carcass. On his next flight Doug saw the Druids at the site where 375 died. They must have followed her scent trail and found her remains, and would have figured out from scent trails at the site that the Sloughs had attacked her.

The pack was down to just six members now: two adult females, two adult males, and two pups. A much larger rival pack was frequently in their territory, had found the Druid den site, and had killed a family member. Strategically, the Druids needed to stay out of the way of the Sloughs.

With the Druids keeping a low profile, the Sloughs were often in Lamar. They killed a cow bison there on the twenty-second. Kirsty Peake, a dog behavioral specialist, was watching two of the young Slough wolves before that successful hunt. She told me that when they approached a lone bison, both of them were kicked so hard they flew backward through the air and crash-landed. The two wolves lay in the snow for two minutes before getting up in what she called a "very groggy" condition.

I finally spotted the Druids on the twenty-fourth in the Chalcedony rendezvous site. All four adults and two pups were in the group. They went east and I lost them going up the Lamar River. Two days later, I got their signals just outside of the park, near my cabin in Silver Gate.

On February 24, seven Agates, including alpha male 113 and the newly collared alpha female 472 came into Lamar Valley. 472 was a former Druid so had grown up here. The Agates sniffed around where the Sloughs had killed the cow bison. Like the Druids, the Agates were greatly outnumbered by the Slough pack. Their alpha male, 113, faced the same issues as 480—for now the wisest strategy for both males was to avoid a fight with the much larger pack. The Agates left Lamar and went back to their territory to the southwest.

In March the Sloughs continued to frequent Lamar Valley and the Druids were often out of sight up the Lamar River. I saw them just once after February 14, at their Chalcedony rendezvous site, and twice got their signals at Silver Gate. The Sloughs repeatedly visited the smaller pack's den site and their fresh scent there would probably cause the Druids to have their pups elsewhere.

THE DRUIDS AND Agates were not the only packs threatened by the Sloughs' territorial expansion. The Geode Creek pack, the group founded by former Druid female 106, was based west of the Slough pack's territory. The Geodes had eleven members, so they were also outnumbered by the Sloughs. In early March, the Geodes were on an elk carcass three miles west of Slough Creek. I saw fourteen Slough wolves charging that way with their tails raised, then lost the rapidly advancing wolves behind a hill. Emily Almberg's friend Mike O'Connell was at a better angle to see what happened next. He told me that the Slough wolves had attacked and killed a gray wolf and that the uncollared Slough alpha female was the primary aggressor. I joined Mike and he showed me the body of the dead wolf. The Sloughs had now killed two wolves in neighboring packs and we suspected they had killed a third.

When the wolves brought from Canada came to the park, each pack was held in separate acclimation pens for two months to get them used to their new surroundings. On March 21, we had a tenth anniversary celebration to mark the release of the wolves back in 1995. That day I saw the Druid and Slough wolves, both of whom were descended from the original packs. The next morning, I saw the first flowers of spring and soon after that spotted a bluebird, welcome sights after the long winter.

Rolf Peterson, the biologist overseeing a long-term study of wolves at Isle Royale National Park, was in Yellowstone at that time. Because that island is heavily forested, wolves are hard to see. Rolf told me that a researcher hiking around the park can go years without spotting a wolf. The park gets about eighteen thousand visitors annually and Rolf said that collectively they have about fifteen to twenty wolf sightings a year. In contrast, the open country in much of Yellowstone meant we could see wolves almost every day. At that time, I was in a stretch where I had seen wolves for nearly six hundred days in a row. The high visibility of our wolves enabled us to see aspects of wolf behavior that had never been well documented before and to study how closely related wolves, such as 302 and 480, can have very different personalities.

Fourteen Slough wolves marched into the Druid den forest once again in late March. We were watching the six Druids a mile to the south, up high on Mount Norris. They howled when they spotted the intruders and the Slough wolves howled back, defiantly announcing their occupation of the other pack's den site. The Druids immediately stopped howling and 480 led them away. It was distressing to see the Druids retreating from what had once been the core of their territory. The Sloughs stayed at the Druid den site and howled for the next two days. The Druids did not howl back, probably because they wanted to conceal their location from the rival pack.

I noticed that 480's former jet-black coat was now graying on his belly and rear end. He was not yet two years old. I wondered if the stress of dealing with the Slough wolves and coping with 302 was aging him faster than normal. Despite the many challenges he faced, 480 often took breaks from his heavy responsibilities and played with the two surviving pups. One morning I saw him roll around on a snow patch, then jump up and run back and forth like a playful pup. The pups saw that, raced over, and chased the big alpha male. He fell to the ground and the pups stood over him, acting like they had just defeated him in battle. A few moments later, 480 ran off and the two pups pursued him. He tripped, probably intentionally, and the pups triumphantly surrounded him and nipped at him. He got up and ran off, his tail held low, while the pups chased him with their tails raised high. Just as I had seen 21 do so many times, 480 was pretending to be a low-ranking wolf as he played with the pups.

On April 1, the Druids went on an elk hunt. They chased a group of bull elk and targeted the slowest one. A wolf grabbed a hind leg and another bit into a shoulder. The bull broke away, but the pack easily caught up with him and the alpha pair bit into each side of his neck. The bull was so tall that both wolves had to stand on their hind legs to maintain their holds. When the bull raised his head, he lifted the Druid alpha female right off the ground. Meanwhile 302 grabbed the bull's right shoulder, and older female 255 and the two pups attacked the hindquarters. The bull tried to run off, but the alphas kept their holds on his neck. The wolves pulled him down and he died a few minutes later. I was happy to see that 302 had been a contributing member of the hunting party.

Both Druid females looked pregnant. We soon learned that the alpha female, 286, was using a new den site up the Lamar River near Cache Creek on the outskirts of their once huge territory. One of the tracking flights spotted seven pups at the site. That was encouraging news. If the Druids had good pup survival, they would be better matched against the Slough wolves in the coming year.

6

Four Mothers at Slough Creek

SINCE THE DRUIDS' new den area was not visible from the road corridor, I concentrated on watching the Slough wolves. One morning in early April, I saw the pack west of Slough Creek, close to a coyote den they had raided in the spring of 2003. Four adult females were there and all appeared to be pregnant. Wolf gestation periods are the same as dogs', about sixty-three days, and we knew from watching breeding ties that the alpha female was due to give birth around April 3.

I saw that there were three burrows dug into an east-facing slope, all close to each other: left, middle, and right. The left tunnel was the old coyote den and the other two looked new. We saw the pregnant females go into the left burrow most often, but they also visited the middle and right dens. Sometimes all four of them were inside the left den at the same time, indicating the pack had enlarged the tunnel since coyotes had originally excavated it. All this activity meant that the pack would be centralized there, many miles from the Druids at Cache Creek.

I heard that a grizzly wandered into the site and looked into the den on the right. It did not stay long for a black female rushed out and drove the bear off. I arrived in time to see her chasing the big grizzly away. The bear did not want to have anything to do with an angry pregnant wolf.

The four adult males and eight younger adult wolves in the pack visited the site regularly to regurgitate meat from their kills to the females. It is much more efficient for a wolf to stuff itself with large amounts of meat at a kill site and convey it to the den in its stomach than it is to carry a piece of meat home in its mouth. When the wolf regurgitates that meat, which looks like chunks of stew meat, it is partially broken down and easier for the pups to chew.

By early April, we were fairly sure that all four females had given birth to their pups in the left den, which we called the Natal Den. The adult males seemed fascinated with that burrow. They would look into it and often go inside to see the young pups.

On April 12, I saw a mother wolf carry a very small pup from one den to another. By April 24, we started to see pups walking around outside the entrance to the Natal Den. They would have been about two and a half weeks old. Big alpha male 490 liked to stand at the entrance and touch noses with each of the pups when they came out. I guessed that sniffing enabled him to memorize the individual scent of each pup. The mother wolves nursed the pups regularly and seemed willing to feed any pups that came to them. I had observed what is known as communal nursing before and learned that it is common in wolves.

That hillside where the dens were located offered a good view of the surrounding country, and the females could spot incoming pack members as well as potential threats, such as coyotes and bears. There was a small stream nearby where the mothers could drink and plenty of elk in the area for the adults to hunt. It seemed like an idyllic setup. One day there was a Disney-like moment when a bull elk went to the Natal Den and looked inside. He must have been hearing the pups and wanted to see what was going on.

By April 24, the four mothers had recovered so well from having pups that all of them went out on a hunt with other adults. The females were in high spirits, probably because they were getting a break from the pups, and they did a lot of playing as they traveled. They engaged in carefree chases, ambushes, and wrestling matches, then returned to the maternal duties that awaited them back at the den.

The open country around the den enabled us to document the details of wolf family life. We watched as wolves picked up errant pups and returned them to the safety of the den. The five yearlings were especially attracted to the pups and seemed happy to babysit them when the mothers were temporarily away, but the younger wolves were a bit inexperienced in some tasks. We saw a gray yearling go into the den and come out holding a black pup by its head. The yearling dropped the pup, then picked it up. One of the mothers came over and grabbed the pup. It slipped out of her mouth, but she nimbly caught it before it hit the ground.

There was a flat area at the den entrance that we called the porch. By April 26, the pups were often hanging out there. The mothers nursed the pups on the porch, giving us a clear view of them suckling. After nursing, the pups often climbed around on the backs of the mother wolves. At first the nursing was mostly done when the mothers were bedded down, but when the pups got a bit older the females often stood during the process, and the pups had to balance on their hind legs to reach a nipple. They would place a front paw against the mother's belly or leg to steady themselves. By May 1, the pups could walk around the den area fairly well. A few days later they were sniffing at the ground and following scent trails throughout the area. They were also doing a lot of playing together, especially wrestling, a favorite game of all wolf pups.

The den area was about a mile from the Slough Creek campground road and crowds of people gathered there every morning and evening to see the pups. My job included helping park visitors see the wolves, so I shared my spotting scope with thousands of people, who were ecstatic when they saw the pups. I was your friendly neighborhood wolf researcher. The site may well have been the best wildlife viewing spot in the world at that time. We got a high count of nine black pups and seven gray pups, but soon the count of blacks dropped to seven.

We saw one of the mother wolves repeatedly digging at an old burrow downhill from the Natal Den in early May. We later determined that it had once belonged to a badger. Another mother helped excavate the site and the alpha female came over to see what they were doing. Soon the females had enlarged the tunnel so much that the big males could walk into it. There was a lot of sage in the area, so it became known as the Sage Den.

The pups gradually switched to that new den. Some were carried over by the mother wolves. Others found their way on their own. Wolves frequently move their pups to new dens and we usually do not know why. Doug Smith told me that many wolf dens he examined were infested with fleas, so that would be one reason to switch pups to a new site. Wolves don't seem to have much difficulty digging new dens when they need to. I once worked with a group of captive wolves. One pair had a litter of pups and when that family was moved to a new enclosure, the two parents worked together to dig a new den. It took them less than an hour.

The yearlings' fascination with the pups was beginning to wane and some of them did not appreciate being harassed when they were trying to sleep. I watched as one tired yearling picked up a pup that wanted to play, carried it off some distance, put it down, then returned to his bedding spot and napped.

The pups instinctively ran into the den if something new or unexpected happened. When the tracking plane flew over, the pups rushed to hide underground, and they did the same thing when they heard thunder. That instinct is known as neophobia and means fear of the new. If a bear, coyote, or human came into the den area when the adult wolves were away, the pups would likely run into one of the dens and hide.

I watched as the adults worked to feed and protect their pups. One day the third-ranking male, 489, returned from a hunt and thirteen pups ran to him, pestering him for a feeding. He lowered his head and regurgitated a big pile of meat. The pups quickly ate all of it. Later a grizzly approached the den area and alpha female 380 got between it and the pups. Three of the mother wolves teamed up and chased the bear away. One of them nipped it on the rear end. The next day a bigger grizzly came into the den area and 489 and other wolves drove it away. Later a bold pup played a trick on 489. As the big male looked into the Sage Den, the pup sneaked up behind him, grabbed his tail, and yanked it several times. 489 backed out and walked off without disciplining the pup. He seemed willing to let the pup have fun at his expense.

The job of babysitting fourteen active, curious pups was not always easy. A male yearling was with a pup who was moving too far from the den area. The yearling hovered over the pup and tried to block it from going farther. A female yearling with a different pup who was straying too far tried to pick it up, but the pup was too heavy for her to carry. When it scampered off, still going the wrong direction, she raced after it and knocked the pup over with a front paw. The two yearlings eventually got both pups to go back to the den. Caring for the pups appeared to be a rite of passage for yearlings. The experience they gained by doing that job would benefit the young adults when they had pups of their own.

I later saw a little pup trying to figure out a challenging situation. It was following one of the mother wolves and she led it to a cluster of logs. The adult easily jumped over a log, but the pup failed at repeated attempts to climb over it. Refusing to give up, it found a slight depression under one of the logs, ducked down, and crawled through it. I was impressed by the pup's persistence, a trait I had noticed in adult wolves, as well. Persistence was a quality they needed if they were to successfully hunt animals that were normally larger and faster than they were.

Cleverness is another quality wolves use to their advantage when hunting. One day I followed 489 through my scope when he went out on a hunt. Seven cow elk were bedded downhill from the den. All were adults in prime condition and unconcerned that a pack of wolves was nearby. 489 slowly moved toward them. The elk got up and the wolf did a play bow to them, like he was daring the elk to chase him. They took the bait and charged at him with a big cow out in front. 489 ran off, but soon looked back over his shoulder. He turned around, leapt up at the oncoming cow, and tried to get a killing bite on her throat. She just barely dodged the attack and ran off.

The trait of cleverness in wolves applies not only to actively hunting, but also to stealing another predator's kill. In May I spotted the Slough alpha male near a grizzly that was feeding on an elk calf carcass. When the wolf approached, the bear chased him off. 490 came right back and nipped the bear on the rear end. It chased him a short distance, then went back to the kill. When the bear stepped away from the carcass, 490 rushed in, grabbed the calf, and ran off with it. Cases like that made me think that wolves were smarter than bears.

HUNTING REQUIRES PERSISTENCE and intelligence, along with good physical coordination. The pups at the den site were getting plenty of practice developing that skill. There was a bow-shaped log downhill from the Natal Den. Each end rested on the ground and the middle section was about four feet in the air. Several big rocks with sharp edges were under that high point. The pups liked to walk along the log in single file. They had good balance, but every time I saw them above the jagged rocks, I was afraid one or more would fall.

I did see a pup slip in that section, but it gracefully regained its balance and continued to the far end, unfazed by its near fall. Another time two pups were balancing on the log when a raven landed between them. The bird, about the same size as the pups, sneaked up on one and pulled its tail. That pup turned to see who had yanked its tail and almost fell off. The raven then went toward the other pup, who was facing away, pecked at its tail, then pulled it. The pup nimbly turned around on the narrow log and lunged at the raven, who flew off at the last moment. The pups constantly played on that log and none ever fell off. Such development of their physical coordination would pay off when those pups were older and fighting rival wolves or tackling prey animals.

I once had one of the captive wolves I worked with play a trick on me like the one the raven played on those pups. I was in the family's pen one morning, crouched down near a young pup. It yipped out in pain, probably because it had been bitten by a bug. I knew the big father wolf was somewhere behind me and worried that he might think I had hurt his pup. Trying to avoid any sudden movements, I began to slowly stand up, intending to turn around to see what he was doing. In the middle of getting up, I felt a light pinch in my rear end. Looking over my shoulder I saw the huge wolf running away. He could have done serious damage to me if he had used the full force of his jaws but decided to just give me a warning. Later I wondered if that incident proved that wolves have a sense of humor. I think that one did.

By late May, when the pups tried to nurse, they usually gave up after a short time. The mother wolves' milk supply must have been drying up. The pups had been eating meat for a few weeks at the time, so they were no longer dependent on their mother's milk. Occasionally, a mother would get into a nursing position, but the pups would lick her face, trying to get a regurgitation from her, rather than attempt to nurse. That indicated they now preferred meat to milk. The pup count continued to be seven black and seven gray pups.

One May morning, when I was hiking back from an observation spot to the road through tall sagebrush, I heard the sound of running animals. I grabbed the bear spray from my belt and got ready to use it. I spotted a mother grizzly and her two yearling cubs running away from me. I realized I had gotten too comfortable walking around that area. From then on when I went through thick brush, I made a lot of noise and had the bear spray in my hand, ready to use. Grizzlies generally want to avoid people and tend to move away if they hear human voices.

Tracking flights throughout May continued to get the Druid signals up the Lamar River in the area of their new den at Cache Creek. The two pups from last year were now yearlings and considered adults. The pack often made kills in Lamar Valley and I regularly caught sight of them there. The wolves would fill their stomachs, then head back to the den where they would regurgitate the meat to the pups. All six adults looked to be in good condition.

7

The Fate of the Slough Pups

THERE WERE TWO groups of trees just north of where the Sloughs were denning. We called one the Diagonal Forest and the other the Horizontal Forest. On June 6, pups were seen going into a burrow behind the Horizontal Forest. We found a spot along Slough Creek Campground Road where we could get a partial view of the new den area. All the surviving pups, seven grays and seven blacks, made the move. We called the new den the Northern Den.

That was the second move the Sloughs had made with their pups that year. They had moved from the original coyote burrows on the slope to the old badger den farther down the slope, and now they were taking up residence in another former coyote den. I hiked up a rocky ridge east of the campground road and found a good viewpoint to monitor the den site from about a mile away. From then on, I went up there nearly every day. The pups spent a lot of time running around and vigorously playing. The adult wolves regularly went out on hunts, then returned to feed the pups.

A few days after the move, a young Slough female was south of the den by herself. A group of three wolves came into the area, saw the gray, and charged at her. She dropped to the ground and went into a submissive posture. One wolf grabbed her throat while the other two nipped at her. The bite on the throat would have been fatal if the wolf had used all the 1,500-pound pressure in its jaws. The Slough wolf thumped her tail on the ground, signaling she was willing to be friendly. She also pawed up at her attacker's face, like a pup would to an adult. The other two wolves lost interest and walked off. The third wolf continued to hold on to the female's throat, then suddenly let go, sparing her life. She jumped up and ran off. None of the bites seemed serious and she soon reunited with her family.

That incident showed how in a conflict with other wolves, a defeated wolf can go submissive, short-circuit an aggressive attack, and survive. It would be like a hiker playing dead when attacked by a grizzly or a mixed martial arts fighter tapping the mat or their opponent to indicate they are submitting when they cannot get out of an opponent's hold. I wondered if the young Druid female the Sloughs caught and attacked had refused to act that way and fought back with all her strength. If so, her defensive behavior probably got her killed.

All three of the young Slough's attackers were collared, and I later learned that there was a backstory to the older male who had spared her life. He had been born into the

Leopold pack. When he was radio-collared, researchers collected a blood sample for analysis. They expected that his father would be the Leopold alpha male, but it turned out to be a dispersing Druid wolf, that is, a wolf who had left the pack and was now looking to join another pack or start a pack of its own. We had gotten signals from him in the Leopold territory during the 2002 mating season, and genetic testing proved that he bred a female there. The gray female whose life was spared would have been born to one of the Druid females that helped to start the Slough pack, so the older male had spared the life of a relative. Maybe he recognized her scent as being like his own and that was why he let her go.

Writing about that incident triggered a memory from 1995, the first summer the wolves were back in Yellowstone. Dogs are allowed in the park as long as they are on a leash and kept near the road. A visiting family told park rangers that their little dog had run off in Lamar Valley and did not come back when called. More than a week went by without any sightings of the dog. We all assumed that it had been killed by predators or had starved to death. Then it showed up next to the road and was perfectly all right. If the small dog had run into wolves, it likely would have acted submissively. It was about the size of a wolf pup, so if it had rolled on the ground and whined, the wolves would probably have classified it as a harmless pup and let it go.

On June 12, I completed five years—1,826 days—of getting up early every morning and going out to study the wolves. That day I got up at 3:20 a.m., the usual time for that time of year. Since I worked every day, I sometimes would forget what day of the week it was and would have to ask someone if it was Saturday or Monday.

ONE DAY I noticed that a gray Slough pup was stretched out downhill from the burrow behind the Horizontal Forest. It seemed to be sleeping, but then some time went by without any movement. Most sleeping wolves shift around a bit and regularly lift their heads to look around. A black pup was walking around near the gray, something that should have caused a reaction, and the gray pup still did not stir.

The pup lay there for two days, meaning it was dead. Then I spotted another gray pup that did not bother to get up when an adult wolf returned to the den. The adult howled and the pup did not react. I saw three nearby black pups that looked active and healthy. They played together and mobbed adults arriving at the den. Then I noticed more gray lumps scattered around and realized they were additional sick or dead pups. Soon magpies landed and began pecking at the lifeless bodies.

All seven of the gray pups died and just three of the seven black pups survived. We eventually determined that the pups died from distemper. The coyote researchers told me they were also finding dead coyote pups at dens. Tests confirmed that distemper was the cause of their deaths as well.

Some of the other Yellowstone packs also lost pups to distemper that year. The Wolf Project later tabulated that pup survival rates throughout the park in 2005 were the worst since the reintroduction in 1995 (23 percent compared with an average of 77 percent in normal years), and the total wolf population dropped by 30 percent.

Dan Stahler was starting a PhD on wolf genetics at the University of California, Los Angeles (UCLA) with Dr. Robert Wayne as his faculty adviser. Researchers in his lab discovered that wolves with black coats have an increased immune response that enabled them to recover from distemper, a naturally occurring virus, at a higher rate than gray wolves. All three surviving Slough pups were black. The research at UCLA also traced black coats in wolves back to a genetic mutation in an early version of the dog, likely after the last Ice Age about ten thousand years ago. Back then all dogs would have looked just like wild wolves since they were only a few generations away from their wild relatives and often interbred with them. The black mutation quickly spread through both wild wolves and dogs. Gray coats come with genetic advantages, as well. For reasons not fully understood, gray mothers have better overall pup survival than black females, even with the periodic outbreaks of distemper. The Yellowstone wolf population tends to be half black and half gray, and we found that our wolves tend to pair off with a mate of the opposite color.

As I continued to watch the Sloughs, it was encouraging to see how active the three surviving pups were. They played vigorously together, ran around the den site by the Horizontal Forest, and pestered adults for feedings. By July they often followed the adults when they left on hunts and later found their way back on their own. They could follow a scent trail home or figure out how to get back through unfamiliar terrain by relying on their sense of direction. Two of the pups were smaller than the third one, and they often teamed up to chase their larger sibling. The big pup was likely a male and the smaller ones his sisters. They seemed to enjoy teasing their brother.

Dan did a flight on July 15 and found the Slough adults scattered throughout their territory. The three biggest males were off on their own, looking for vulnerable prey, a hard job in midsummer when adult elk are in peak physical condition and their calves are old enough and strong enough to outrun wolves. In addition, most elk were now far from the den area in high-elevation meadows where there was fresher, more nutritious grass for them to eat than down in the valley. That evening the Slough pups traveled several miles from the den with the adults and easily kept up with them.

Doug sent a crew to examine the three Slough dens later that year: the Natal Den, the Sage Den, and the Northern Den. The Sage Den had a lot of fleas, but there were none at the Northern Den, which suggests the mothers might have made that last move with their pups because of a flea infestation. The research crew found scattered pup bones at the Northern Den, including small skulls and jaws from the ones that had died at the site. Emily Almberg told me the teeth in the jaws were discolored, a sign of distemper.

After the loss of thirteen of the sixteen pups in the Slough pack, I thought a lot about how people who study wild animals cope with such incidents. To do this work you have to be able to deal with the reality of life in the wild. The Yellowstone wolves do not live in a fairy-tale world where nothing bad happens. My purpose was to observe and to understand what the lives of wild wolves are really like, the good times and the difficult times, then tell their stories to people so they can know what it might be like to be a wolf.

Each pack must control a territory with enough prey animals to support the family and will likely have to fight rival packs to retain control of that land. In those battles wolves from either side could be seriously injured or killed. Hunting large prey animals like elk and bison is a dangerous occupation and wolves regularly get hurt on those hunts, sometimes fatally. Although wolves have the potential to live as long as dogs, in Yellowstone the typical life span is only four to five years.

In 2009 I learned from Dan that the alpha male in the Bechler pack in the southwestern corner of the park died just after his twelfth birthday. A twelve-year-old wolf would be equivalent to a person at ninety-three, a huge achievement for a wild animal, especially for an alpha who had to continually battle prey animals much bigger and stronger than himself. Years later an alpha female in the Cougar Creek pack lived to be the oldest known wolf in Yellowstone: twelve and a half. She had pups when she was eleven.

Despite all the challenges wolves face and their often short lives, they seem to love life and get great joy from being with their families. I imagine that they would never want to be anything other than a wolf.

I had been getting periodic reports on former Druid male 253 for several months. In January I heard from Mike Jimenez, a U.S. Fish and Wildlife Service biologist stationed in Wyoming, that he had been seeing 253 in the National Elk Refuge, near Jackson. That site is south of where Mike released him after he was trapped in Utah in late 2003.

Soon after that, two wolves were seen traveling with 253 at the Elk Refuge: a gray female and a black male who acted subordinate to 253. In mid-May, I heard that 253 and his pack were denning on the Elk Refuge. Soon after that, six pups were seen at the den. The black male turned out to be Druid 350, born to 21 and 42 in 2003, so he would be 253's younger brother. 253's group was designated the Flat Creek pack.

Later there was a report that two people and a dog unwittingly walked into the Flat Creek wolves' rendezvous site when the pups were there. A large gray wolf escorted them out of the area and did a lot of barking and howling at them. Since 253 and 350 were black, that gray was probably the mother wolf and 253's mate.

After that I got word that there were now four adults in 253's pack, including the gray alpha female, Druid black male 350, and one other black adult. If the six pups survived, the pack would number ten. With the loss of so many pups in the Slough pack, it was good news to hear how well 253 and his family were doing.

8

The Sloughs Expand Their Territory

LATE IN AUGUST, the three surviving Slough pups were with the adults in Lamar Valley at an elk carcass. Three of the older wolves had left the family, probably to find mates, so the pack now had twelve adults. The three pups were about two-thirds the height of the adults and looked to be in good condition.

Now that the pups were older, the pack was based at a rendezvous site at Jasper Bench, at the west end of Lamar Valley. The pups mostly stayed at Jasper Bench, but the adults often traveled east and were frequently seen at the Druids' Chalcedony rendezvous site, the Druid den forest north of the Footbridge parking lot, and areas as far east as the park border near my cabin in Silver Gate. All that had once been Druid territory, but the Slough wolves had taken it over. They went wherever they wanted, just like the Druids in the early 2000s when they were a superpack.

A lot of elk and bison traveled through Jasper Bench, and the Sloughs frequently made kills in the area. One day a bull elk ran down off the bench into the river and the wolves finished him off in the water. By that time, the pups had graduated from eating regurgitated meat brought to them by the adults to feeding directly on carcasses, so the pack camped out by the carcass for a few days. I watched as one of the pups waded out toward the elk, stopped to look down into the water, then jumped up and pounced at something, probably a trout. The pup missed.

Like all young wolves, the Slough pups were good at finding things that were fun to investigate and new kinds of food. They found a burrow on Jasper Bench, likely an abandoned coyote or badger den. They liked to dig at the entrance and go in and out of the tunnel. There was a high population of grasshoppers that summer, and the pups would catch and eat them.

In mid-September, when all the Slough adults were off on a hunt, the three pups went on an expedition and found the remains of a bison that had died seventy-three days earlier. They grabbed bones from the site and gnawed on them, like a pet dog would. The jaw strength in those pups was not sufficient to crack open those big bison bones and gain access to the marrow, but adult wolves can do that.

By the end of the month, the Slough pups were often with the twelve adults at the Druids' old rendezvous site at Chalcedony Creek. The three pups found the marsh where in past years I had seen Druid pups hunting for voles, and they were soon pouncing on the little rodents. The Slough wolves continued to be based there into October.

In addition to the usual elk and bison, we were now occasionally seeing moose in Lamar Valley. Moose are browsing animals and feed on willows along the river and streams. One morning a bull moose walked by the bedded Slough wolves and they ignored him, probably because they were unfamiliar with moose and possibly intimidated by his size and huge antlers. In places like Isle Royale National Park in Michigan, moose are the primary prey for wolves. In Yellowstone, moose make up only about 1 percent of their kills.

WE HAD NOT seen the Druids much that summer. On August 25, we spotted 480 and 302 with two female yearlings and one pup at the Chalcedony Creek rendezvous site. We wondered if they were checking it out to see if the Sloughs had been there recently. Alpha female 286 was not with them. The battery on her collar was dead, so we could not track her. Older female 255's collar was giving out a mortality signal from a spot south of Cache Creek, which meant she had either died there or dropped her collar at that site.

When Doug Smith flew on September 7, he got signals from 302 and 480 way up Miller Creek, a tributary of the Lamar River, twenty miles from Lamar Valley and the Slough wolves. A flight in mid-October spotted the same five Druids that I had seen in the rendezvous site. We reluctantly concluded that 255 and 286 had both died, possibly from distemper. The high count of pups had been seven but there was only one survivor. These deaths were a severe blow to the Druid pack's chances of recovering their territory in Lamar.

Things were not to improve. Later that month, we saw the Druids feeding on a bull elk carcass and noticed that the pack was down to just four: 480, 302, and two female yearlings. The last surviving pup had disappeared. None of the lost Druid pups were ever found, but we suspected that at least some of them had died of distemper, like so many of the Slough pups had.

When the four Druids had eaten their fill, they moved off to the west. Suddenly they stopped and intently looked east. Emily Almberg called and told me the Sloughs were howling. The Druids' actions indicated they were hearing the howls. We estimated that the two packs were eight miles apart. The Sloughs continued to do more howling as the Druids silently slipped away to the west. I found them at Slough Creek the next morning with a fresh elk carcass. The Slough pack had invaded the Druid territory. Now, the Druids had gone into Slough territory and killed an elk, a defiant response to what their rivals had been doing to them.

After watching the Slough wolves and their pups so much that year, I realized that they were just like any other wolf pack. They looked for opportunities to enlarge their territory and tried to outcompete rival neighboring packs. For now, they outnumbered the Druids and could push them around. In earlier times, the Druids had done the same thing to their rivals. They had many years of dominance and now were going through a period of rebuilding. It was the normal ebb and flow that wolf packs had been going through for thousands of years. The Sloughs were just being wolves.

The three big males then in the Slough pack—490, 377, and 489—and the founding alpha male, 261, had all grown up in the Mollie's pack. That family was originally known as the Crystal Creek pack. They were brought down from Canada in 1995 and were the first wolves to colonize Lamar Valley. The Druids attacked them in the spring of 1996 and killed their alpha male. His mate and the only surviving male had to abandon Lamar and settled into a new territory in Pelican Valley, about twenty miles to the south, where they thrived. Now, nine years later, the descendants of the original Crystal Creek pack once again controlled Lamar Valley and had their multigenerational rivals, the Druids, on the run. It was like a story out of the Old Testament or *Game of Thrones*.

FALL WAS ENDING and temperatures were dropping. We get a lot of snow where I live in Silver Gate, at an elevation of 7,389 feet. The average yearly snowfall is 169.5 inches, just over 14 feet. We normally have only two months per year without any snow: July and August. One morning in early December I got stuck in heavy snow in my driveway and had to dig my way out. The next morning it was minus 32 degrees Fahrenheit (–36 Celsius). I tried to leave, but my battery was dead. I jump-started the car with a portable battery and had to drive out of the park to buy a new battery. The low that day in Lamar was minus 40 degrees. There was no wind, so the extreme cold was tolerable as long as I went back to my car to warm up. During the early years of the Wolf Project we had a motto: "Get the data, regardless of the cost." I was willing to pay the cost.

The wolves went about their lives, seemingly immune to the cold. That month Emily saw eight of the Slough wolves chase a skunk. It sprayed some of the pups who got too close to its business end. They tried to wipe off the stink from their faces with their front paws. A more experienced adult killed it. She ran off with it and the pups chased her, hoping to steal the skunk.

I spotted the four Druids later in December. Despite all the hard times they were enduring, they spent a lot of time playing together. The two older males played with the yearling females like they were all pups. They chased each other and wrestled as they rolled around in the powder snow. One male ran down a slope and seemed to deliberately do a pratfall, just for the fun of it.

A few days later, I watched the Slough wolves struggling to travel through deep soft snow. The big second-ranking male, 377, was leading. To make progress he had to bound up and down, which was exhausting. After a short distance, he collapsed and rested. When his strength returned, he repeated the bounding but had to rest once more. Soon extreme fatigue forced him to rest for longer periods. Impatient with his slow progress, a younger wolf jumped over him and took over the lead position. Later other wolves took their turn breaking trail. That is one of the advantages of being in a large wolf pack: the members can trade off on exhausting jobs like that.

On Christmas Day, I saw the first signs of the coming mating season. A black Slough female tried to flirt with 377. He sniffed her and apparently decided that she was nowhere near being ready to breed, so he walked away. Later alpha male 490 pinned 377 and then did the same to 489, reminding them of their place in the male hierarchy.

PART 3

—◆—

2006

9

The Mating Season

A T THE START of 2006, wolf 302 was over five and a half years old and had already lived beyond the average life span for a Yellowstone wolf. He was still a subordinate male in the Druid pack to his younger nephew and appeared to have no prospects of becoming an alpha. 2005 had been a tough year for both his family and the Slough wolves. The Druids lost their alpha female, 286; their only other remaining adult female, 255; and all six pups. They were down to just four members: 480, 302, and two female yearlings. Their rivals, the Slough wolves, had had four litters with at least sixteen pups but distemper had killed all but three of them. The pack now numbered twelve adults and the three surviving pups. 2006 would be a far worse year for one of those packs.

I watched the four Druid wolves on January 8 at a new kill. 480 carried off the vertebral column and bedded down. The spine still had a lot of ribs attached and both yearling females, a black and a gray, came over and all three wolves chewed on different sections of the column at the same time. Later the two sisters played together. The black did a flexed-leg urination, a sign she was the dominant female.

The male hierarchy in the Slough pack remained as it had been in 2005: the alpha male was 490, the beta male was 377, and 489 was the third-ranking male. 380 was still alpha female. Doug Smith collared two female Slough wolves that month: yearling 526 and adult 527. He then went on to the Druids and collared the black yearling, the pack's new alpha female. She would now be known as wolf 529. Later in the year, the other Druid female was collared and designated 569.

I did not see any wolves on January 22, 2006. Up until then, I had seen wolves in Yellowstone every day for 892 days in a row.

In the Yellowstone area, wolves mate from late January through much of February and the pups are usually born in April. By the time the litters are weaned, about five or six weeks later, elk are calving—a pulse of life that usually enables wolf parents to keep their pups well fed.

We saw the Slough alphas mating on January 26. After that the gray Druid yearling female flirted with 302, and 480 made several attempts to mate with the alpha female, 529. He repeatedly tried to keep 302 away from 529. At one point, 480 lunged at his uncle and 302 fell over submissively. When 302 later approached the alpha female, 480 pounced on him. Right after that, 480 rushed over to 529 and they mated. After they finished, she went over to 302 and bedded down next to him. 480 ran over and got between them. Despite 302's failure to be a proper alpha male, the females were still drawn to him rather than the hard-working 480. Laurie Lyman, my Silver Gate neighbor and a longtime wolf watcher, noticed that 480 often bedded down near 302. That enabled 480 to closely watch the older male and intervene if he tried to approach one of the females.

Laurie was very skilled at spotting wolves, telling individuals apart, and picking up on subtle aspects of their behavior. She also had a good memory of how individual wolves had behaved in the past. When I asked her about her abilities, Laurie attributed them to the many years she taught in an elementary school. Learning how to quickly tell all the kids apart when classes started each fall and watching their behavior on the playground enabled her to easily identify wolves and discern their personality traits. She was a big help to me and the Wolf Project.

480 bred 529 again on February 8, and while they were tied, 302 mated with the gray yearling, 569, just twenty feet from the alpha pair. 480 saw what they were doing but was still tied to 529. Wolves lock together at the beginning of a mating tie and cannot break free for several minutes. 480 dragged 529 to the other pair and bit at 302, but it was too late to stop the breeding. 302 yipped out in pain every time he was bitten. 529 did not do anything to her sister while that was going on. 302's actions were an example of how he could get away with something by waiting for the right moment to make his move. It was like the time he bred the previous Druid alpha female as she was approaching the park road and 480 was hesitating farther behind them.

When I was writing this section, I consulted with Dan Stahler about the genetics of the eleven pups born in the

Druid pack later that year. 569, the female we saw mating with 302, must have also been bred by 480. Seven of her pups were later collared and their DNA showed that 302 was the sire of one and 480 had fathered the other six. We know that in dogs there can be multiple sires in one litter. This was the first documented case of that happening with wild wolves.

10

The Den Siege

I N LATE MARCH, the Druid signals came from their den forest near the Footbridge parking lot. I knew that the Slough wolves had been there recently. They had not stayed long, but the Druids would have picked up their scent. I felt that meant the Druids were unlikely to use their traditional den site this year.

I guessed that the Druids would go back to their 2005 den site near Cache Creek—instead, I frequently got their signals in the Round Prairie area. That meadow lies along the park road five miles east of the Druids' traditional den site. Pebble Creek and Soda Butte Creek meet there, creating a lush habitat that serves as a prime feeding area for elk, bison, and moose. Thickly forested slopes rise up on all sides, and mountain goats live in the rugged cliffs above the trees. As the Druids were spending a lot of time there, I took that to mean they were going to den somewhere nearby, well away from the Sloughs.

The Slough wolves had visited their denning area at Slough Creek in the second week of March. That indicated they intended to den there once again, despite losing most of their pups to distemper in the burrow behind the Horizontal Forest the previous year.

On April 1, three female Slough wolves, alpha 380, 526, and an uncollared gray, ran to alpha male 490 and pestered him for a regurgitation. He brought up a big pile of fresh meat and the females gulped it all down immediately, which likely meant that the three females were pregnant. The uncollared gray had a sharp right twist in her tail, so we called her Sharp Right. Since I had seen the second-ranking female, 527, mate with beta male 377, she was likely pregnant as well. Four litters would be a lot of pups for the adults to care for.

Four days later, I heard wolves barking from behind a ridge to the east of the Slough den site. Slough alpha female 380 was at the den area and barked back. Wolves, like dogs, bark to warn pack members of danger. I hiked up a ridge and spotted twelve wolves, six adults and six yearlings, bedded down a few miles east of Slough Creek Campground Road. I did not recognize any of them. We suspected they had come in from north of the park. One of the gray males had a radio collar but I could not pick up a signal from it, meaning the batteries were likely dead. The pack howled. I got a report that the Slough wolves howled back. The two packs continued howling at each other into the next day. Since we did not know their origin, we called the wolves the Unknown pack.

Dan Stahler did a flight on April 8 and called down on his radio to tell me he got a mortality signal from the Sloughs' third-ranking male, 489, up Slough Creek. He spotted blood in the area and saw ravens pecking at the wolf's body. A Wolf Project crew hiked out there and estimated the date of his death to be April 5 or 6. It looked like wolves had killed him. 489's body was found just a few miles north of where I had seen the Unknown wolves, so they were likely the ones responsible. A gray adult female he often traveled with was not seen after that, and we suspected that she was also killed by the invading wolves. With the deaths of these two wolves, the Sloughs now numbered ten adults and the three pups from the previous spring.

By April 12, it looked like at least three collared Slough females—380, 526, and 527—were secluded in the Natal Den. Sharp Right was probably in there as well. In the early evening, other Sloughs killed a bull elk east of the den. The Unknown wolves were seen traveling west toward Slough Creek later that evening. If they continued in that direction, they would get the scent of the fresh kill, run toward it, and encounter the Slough wolves. It was getting too dark to see, so I had to head in. I figured there would be a battle between the two packs that night.

I got to Slough Creek early the next morning, set up my spotting scope, and looked at the den area. As was normal, many of the adults were bedded down below the Natal Den entrance. I counted five adults and four yearlings. There were only three yearlings in the Slough pack. That tipped me off as to what was happening.

These were not the Sloughs, but the Unknown wolves. They were camped out within a few hundred feet of where the Slough females were hiding in the Natal Den. Some of the intruders left to feed on the bull elk the Sloughs had killed the previous evening. A black adult with a thin white blaze on her chest did a scent mark, which indicated she was the alpha female. The gray male with the collar marked her site, identifying him as the alpha male. Both howled. A black yearling's right ear hung down from its normal position and flopped back and forth as the wolf traveled. That was a sign the two packs had fought during the night. The Unknown wolves must have won.

I got out my telemetry equipment and got signals from Slough alpha male 490 and beta male 377 to the west. I did not get any signals from the three collared mother wolves, which likely meant they were in the den. One of the Unknown adults went up to the den. More wolves joined it, and some looked in. Then the wolves moved off and sniffed around the area.

I did another check and continued to pick up signals from the two Slough males to the west. Then I heard that they were seen heading farther away from their den. That surprised me for I thought that they would go to the den area to check on the pack's females. But if the Sloughs had been badly defeated during the night, perhaps the two males were afraid to go back there. If they encountered the rival pack it would be two wolves against nine.

The next morning, April 14, I got good signals from Slough alpha male 490 and spotted him bedded down alone on a rocky ridge a mile north of the den. He was looking toward that area. Nine Unknown wolves were close to his pack's den site. It was sad to see this big, tough alpha male acting like he was too fearful to do anything to help his family.

I guessed if the Slough females were still hiding in the den, their signals would be blocked from my angle. I walked down Slough Creek Campground Road to a spot where the den entrance was pointing my way and did another check. I got weak signals from 380 and 527, which is what you would expect if they were at the back of the den tunnel. Later I also picked up a signal from 526.

That evening I got weak signals from alpha male 490 to the north and the same poor-quality signals from the three collared females in the den. Based on the mating ties I had seen involving alpha female 380 and female 527, I estimated that their pups had likely been born sometime between April 1 and April 10. Since I was getting signals from female 526 from the den, she probably had pups there as well. If the uncollared Sharp Right was in the den, there could be four litters of pups in the tunnel.

The Unknown wolves stayed at that site for the next few days. They frequently ran up to the den entrance and tried to get inside but were always blocked, apparently by the females in the tunnel. Their siege of the Slough den was like medieval sieges of castles, and just like the occupants of those castles, the occupants of the Slough den would eventually run out of food. The mother wolves had no water in the den and would have already eaten any meat they had taken inside before the siege.

The Slough males had shown no signs of trying to rescue the females by driving off the Unknown wolves, nor had we seen them try to slip past them to bring food to the den. I had spotted a gray two-year-old known as Slight Right because of the slight bend in his tail moving toward the den.

But after looking in the direction of the Unknown wolves, he turned and retreated. The Sloughs were down to three adult males—490, 377, and Slight Right. For now they were separated from each other and disorganized, leaving the females on their own. Soon after that, we rarely saw Slight Right and eventually lost track of him.

Water was more critical to the mothers than food because they needed large quantities to support their milk production. There was a small stream just downhill from the den and it would normally take a mother only a few minutes to go there, drink her fill, and then go back to her pups. But that route would be in plain sight to the Unknown wolves who would see her, charge down, and attack.

On the third day of the den siege, the Unknown wolves were bedded down below the den. The big collared male went to the entrance and stuck his head into the tunnel. Suddenly he turned around and moved off in a rush, like a wolf inside the tunnel had bitten him.

When the Unknown wolves were temporarily away from the den that evening, I saw an uncollared black Slough wolf come out of the Natal Den, look around, then duck back into the tunnel. I got signals from all three Slough collared females toward the den opening. That meant that they were holed up there, along with that black who would have been a yearling. If Sharp Right was in there as well, that would add up to five adult wolves in the den in addition to the pups.

On the fourth day of the siege, a black female from the Unknown pack looked in the den, then went all the way in. She soon backed out and I noticed that her belly was sticking out on both sides, a sign she was pregnant. Ten more

Unknown wolves arrived, bringing the count to eleven, and they clustered around the den entrance.

I watched one wolf go into the den, then immediately back out and run off with its tail tucked. Like the Unknown alpha male the day before, it must have been bitten by an angry mother wolf. Weeks later, after the siege ended and all the wolves had left the area, Doug Smith sent a crew up to the site. The den went twenty feet into the hillside. There was an enlarged chamber at the back end where the pups would have been born and nursed. The tunnel was barely wide enough for one adult wolf to move through, which meant a Slough wolf could stand inside the entrance and bite at any Unknown wolf trying to enter. Thanks to the narrow passageway, a single wolf could block an entire rival pack. That reminded me of the three hundred Spartans, who in 480 BCE tried to block fifteen hundred Persian soldiers from getting through Thermopylae Pass and marching on to invade Greece. The narrow section of the Slough den was equivalent to that pass.

So far the Slough females were successfully defending their den and pups from the rival wolves, but lack of food and water would take a toll on them. That morning I got strong signals from Slough males 490 and 377, but I did not see them. Later their weakening signals indicated they were leaving the area. It looked like they were giving up on trying to help their females and newborn pups.

During the time of the Slough den siege, I also kept track of what the Druids, who were about fourteen miles to the east, were doing. Early in the morning of April 17, I spotted 302 in Round Prairie. He went out of sight at the southeast side of the meadow. That suggested the Druid den was in that direction.

It was foggy when I got to Slough Creek later that morning, and I could not see the den clearly. At times the signals from females 380 and 527 were loud, meaning they were out of the den. 526's signal was weaker, so she was probably inside the den with the pups. Signals from Slough males 490 and 377 were both loud to the north.

When the fog finally cleared, I saw 527 walking around below the den. There was no sign of the Unknown wolves. Then a black Slough yearling came out of the den and sat up. After she went back inside, another black came out and soon slipped back in. I stayed until noon. When I returned at four thirty that afternoon, twelve Unknowns were back at the site. I hoped the Slough females had finally gotten some food and water while the invading wolves were away.

527 was bedded down by the den entrance the next morning. The Unknown wolves were once again gone, probably on a hunt. I spotted a fresh elk carcass a few hundred yards from the den. The Slough females must have come out of the tunnel, killed that elk, stuffed themselves, then returned to their pups, probably carrying big pieces of meat. If their males were not going to help them, they would have to do everything as a sisterhood. That thought reminded me of the 1980s song "Sisters Are Doin' It for Themselves" by the Eurythmics and Aretha Franklin.

Alpha female 380 then came out of the den. There was a snow patch nearby and she ate thirty-five mouthfuls of snow to rehydrate from nursing. She looked around, then slipped back into the den. Right after that, the Unknown wolves returned. The pack went up to the den and one of them sniffed where 380 had been eating snow.

The siege had gone on for five days now and the periodic departures of the Unknown wolves were giving the Slough mother wolves a chance to survive their ordeal. As far as I knew, no one had ever witnessed a situation like this, so several of us organized to make sure there would be good observers at the Slough den site from first to last light every day. Some were with the Wolf Project and others were local wildlife guides and veteran wolf watchers. We shared whatever observations we had with each other.

The alpha female of the Unknown pack had no pups that spring. We did not know what might have been her problem. I felt that was why her family was outside of their territory when they should have been based at their own den, far from here.

That evening the Unknown wolves were still at the Slough den area. A Slough black yearling was feeding at the new carcass. She cautiously moved back toward the den and ended up on a ledge above the site. That gave her a good view of the bedded Unknown wolves spread out downhill from the den entrance. None of them noticed the yearling. She moved down the slope, took one last look at the other wolves, then slipped into the den where she likely regurgitated meat to the mother wolves. That was an extraordinary act of bravery that could have gotten the young black killed, but she pulled the mission off.

She got into the den just in time for a few moments later the Unknown wolves ran uphill and crowded around the entrance. One of them must have spotted the yearling going into the den. Some of the wolves went partway in, then backed out. They apparently could not get past the Slough sister guarding the tunnel from the inside.

Later I saw the Unknown black female, the one who looked pregnant, go into the Sloughs' nearby Sage Den. She spent six minutes inside, then came out. I had already seen that she was a low-ranking female, which made her situation complicated. Since the pack's alpha female was not denning, she was free to roam far and wide, but this wolf was in a bind. She needed both a den and a lot of support from her family. If the Unknown wolves were going to camp out at Slough Creek, she had no choice but to have her pups here. That in turn probably meant that this was not going to be just a temporary siege by the Unknown wolves but a permanent occupation of the territory. This was a nightmare scenario for the Slough pack.

On the nineteenth, I spotted Slough males 490 and 377 in Lamar Valley with a gray female we did not recognize. 490 did a scent mark and the gray marked the site. The pair continued to do double scent marks. That indicated the gray was taking on the role of the alpha female in this new group. The four Slough mother wolves, along with the younger adult females who were helping them and an unknown number of newborn pups, were trying to survive against overwhelming odds. Their two males not only had abandoned them but had taken up with a new female.

The behavior of the Slough males astounded me, especially compared with other alpha male wolves I had known. If this had happened to 21's family, there was no doubt in my mind that he would have gathered up as many pack members as he could, then charged at the Unknowns. The Slough den was located within a mile of the location where 21's adopted father, wolf 8, had fought and defeated the much larger Druid alpha male 38 when he and his pack threatened 8's family in the spring of 1996. After having known alpha males 8 and 21 and the way they fearlessly fought to defend their families, I could not understand the failure of the Slough males to do anything significant to help their desperate females and pups. But then I thought about 302 and how he was so different from his father, Leopold alpha wolf 2, and his nephew, wolf 480, who were both classic alpha males.

THOSE DIFFERENCES IN behavior caused me to think a lot about personality and character in both people and wolves. It seems to me that personality traits are tendencies we are born with such as the tendency to be shy or bold, to be sociable or to be a loner. Some shy wolves find it hard to cross a road if any cars are in sight, even if the cars are some distance away. Bolder ones look for a break in traffic and get across then.

Character traits involve choices. Does a tired wolf choose to get up and go on a hunt to feed its hungry family? Will a young inexperienced wolf take part in an attack on a large elk or bison that is vigorously fighting back, or will the young wolf stand back because it is afraid to get hurt? If a wolf sees a packmate being attacked by rival wolves, will it choose to run in and help or will it run away to save itself? The harder choice is the one that benefits the group but is difficult or risky for the individual.

I once was talking to a young woman who was watching wolves with me. She told me that in elementary school she had been mean to a girl who had no friends. One day she saw a popular classmate go out of her way to be nice to that girl. From then on, she chose to act differently toward other people, especially to the girl who did not have any friends. She saw her popular classmate's action and thought, "I want to be like that!"

I remembered when 302 was the Druid alpha male in 2004 and got into a fight with the big Mollie's male 379. Instead of defending his family, 302 abandoned his pack to save himself. The next morning, 379 came back and fought with the younger 480. Even though 480 was losing, he chose not to give up. Gradually he managed to bite his opponent more often than he was getting bitten. Finally, the big Mollie's male had enough and ran away. Since 302 and 480 were closely related and grew up in the same family, their different characters suggest their behavior was not set by their genes but by their choices.

A kinder way to think of 302 was to compare him to Jeff Bridges' slacker character, the Dude, in the movie *The Big Lebowski*. He was a lovable guy, but he avoided responsibility and always took the easy way out. You could not depend on him for anything. That was certainly how most of the wolf watchers preferred to view 302: as a lovable slacker. One day the Druids were resting on a snowy slope. 302 shifted his weight slightly and ended up sliding down the slope, looking like a hapless character in a movie comedy. That incident only seemed to further endear him to the wolf watchers.

I was impressed with other aspects of 302's personality, however, such as his willingness to try new things. On one of the many trips he made from his family's territory to visit the

Druid wolves, he was seen crossing a bridge over the flooded Yellowstone River. Few wolves could figure out how to use a bridge, but 302 did. He was also an expert in stealing meat from bears. One day I spotted him bedded near an elk carcass in Lamar. A grizzly was feeding on the elk and another bear was walking away from the site. People at the scene told me that the two grizzlies had fought over the carcass. During the battle, 302 ran in and ate as much meat as he could until the bear that won the fight came back.

I continued to monitor the ongoing saga of the Slough wolves and spotted males 490 and 377 on a new kill in Lamar Valley on April 21. The gray female was with them. The Unknown wolves were still at the Slough den on this, the ninth day of the siege. I saw female 527 standing defiantly at the den entrances and howling. She was probably howling to the two males in a futile attempt to get them to come back and help them.

The rival pack was just downhill from her and none of them went after 527. They had probably learned that the Slough wolves could instantly duck into the den and defend it from entry. 527 bedded down just inside the den and looked out like she was hoping to see 490 and 377 coming back. She frequently turned her head to look into the den. The pups were likely crying out to be nursed. It must have been a terrible feeling for a mother wolf to hear a sound like that and know they were unlikely to survive.

The next morning, I saw the Unknown pack's alpha male at the den entrance. He looked in and then immediately jumped back. This was a big male, but he seemed afraid of whatever was in that den. I had great respect and admiration for those Slough females heroically protecting their pups without any help, especially for the one that scared off that alpha male.

I later saw 380 and 527 slipping out of the den and going uphill, probably desperate to find food. They often stopped and looked back to where ten Unknown wolves were sleeping. Suddenly both females ran down the slope toward the den. I saw the rival wolves running uphill directly at them.

The lead Slough wolf reached the den entrance but seemed to have difficulty squeezing in the entrance. Probably another packmate was just inside the tunnel, guarding the entry. The other Slough female arrived and had to stand there, waiting for the first one to get all the way in. The rival wolves were a few moments from reaching them. The lead Slough wolf finally slipped into the den and the other one got inside just as the Unknown wolves arrived. They excitedly crowded around the den and tried to get in, but the wolf guarding the tunnel must have blocked them for none made it inside.

On the twenty-third, I spotted the two Slough males in Lamar with the two-year-old Slough female Sharp Right and a Slough black yearling. The gray female was there as well but she was keeping her distance. I went to Slough Creek and saw 527 and two black yearlings in the den area. That accounted for all three of the surviving pups from the previous year.

The next day, the pregnant Unknown wolf went into the Sage Den and was still inside it when I left three hours later. She probably was having her pups that day. That was bad news for the Slough wolves for it indicated that the rival wolves would be staying a long time at Slough Creek. That meant the Slough wolves were now essentially homeless.

April 25 was the thirteenth day of the siege. I spotted the Unknown female come out of the Sage Den, then go back in. Later Slough female 526 went to the Sage Den, sniffed around nervously, then walked away. The Unknown female soon came out to check where 526 had been. By then 526 was back at the entrance to her own den. She looked at the other wolf and both stared at one another. The two mothers seemed to have an understanding that they would leave each other alone.

Later that day, I got signals from the two Slough males and females 380 and 527 at the unoccupied Druid den, the one they had abandoned that year because of the presence of the Sloughs. I heard the four Sloughs howling from there. Having two of the Slough mothers there was a bad sign. It probably meant that their pups had died and they had given up on protecting their besieged den.

On April 26, the Unknown wolves were still at the Slough den area. A gray yearling from that pack went into the Sloughs' Natal Den, stayed inside for two minutes, then came out. Later a black Unknown yearling repeatedly went in and out. There was no indication of any Slough signals in the area. This was evidence that all the pups had been lost.

Having abandoned the Natal Den, the Sloughs were now gathering in Lamar Valley, about eight miles from the Unknown wolves. Laurie Lyman spotted seven Slough wolves in the Druid den area: the alpha pair 490 and 380, the second-ranking male 377, female 527, Sharp Right, and two black yearlings. She also saw the gray female who had been hanging out with the males in the area. The newly arrived Slough females must have driven her off. The next day, the seven Sloughs were on the south side of Lamar Valley on a kill. Missing fur on the undersides of 380 and 527 and distended nipples on 380 were strong indications that all three had given birth to pups. That seemed to end the siege story, but one more tragedy was to strike the Slough wolves.

Over at Slough Creek, the invading force of Unknown wolves were now firmly in control of what had been the Slough pack's territory. For now the surviving Slough wolves were based in Lamar Valley and could be there permanently. That was very bad news for the Druids.

ON THE MORNING of April 28, I got signals from the two adult Slough males at Slough Creek. We spotted 377 sitting below the west bank of Slough Creek where he seemed to be trying to hide. He looked to be severely injured. Seven Unknown wolves were traveling away from him in the direction of the Sage Den. When they got farther away, 377 went to the creek and drank a lot of water. I saw blood on his right side, both of his hips, and three of his legs. He seemed too injured to get back up the bank, so he bedded down under a log, once again like he was hiding. I spotted 490 to the south. He also seemed injured and was walking very slowly, limping on a hind leg. He howled as he looked toward Slough Creek, probably trying to contact 377.

By that time, 377 had moved up to the top of the bank. His head was down, and he did not howl back, likely not wanting to give away his location to the Unknowns. He went back to the creek three times and got long drinks on each trip. That evening 377 was once again hiding under the log. Early the next morning I got a mortality signal from him. He must have died during the night from blood loss and shock.

The following morning, the last day of April, I spotted 490 with other Slough wolves in Lamar. His front right leg was wounded. I took that to mean that he and 377 had gone back to Slough Creek to see if the Unknown wolves were still at the Natal Den and were attacked. He survived but his brother did not.

The events that spring were a devastating blow to the Slough wolves. Two of their three big males, first 489 and now 377, had been killed by the Unknown wolves and all their pups were lost, despite heroic attempts by the mother wolves to save them during the siege. All that was on top of what had happened the previous denning season when only three of their sixteen pups survived. The pack was now down to eight: the alpha pair, 490 and 380, and six younger adults. We lost track of the gray female that 377 and 490 had been with during the siege and assumed the Slough females had driven her off.

There is an old saying: The enemy of my enemy is my best friend. For the Druid wolves, that meant that the Unknown pack had given them a chance to stage a comeback and regain the territory they had lost to the Sloughs.

The Druids had only four adults, compared with eight in the Slough pack, but both Druid females likely had pups at their new den near Round Prairie. If the pups had a good survival rate, the Druid pack could outnumber the Sloughs at the end of the year.

There is one more thing to mention about the 2006 den-ning season. Early in April, I had a sighting that was not especially significant at the time. I saw Agate alpha female 472 and she looked pregnant. One of the tiny embryos she was carrying that day was of a female pup who would grow up to play a major role in the Yellowstone wolf story and become famous worldwide through a television documen-tary made by Bob Landis. That pup would be known as the 06 Female after the year of her birth.

11

Resurgence

WE SAW THE four Druid adults chasing elk in Round Prairie on May 5. The new alpha female, 529, and female 569 both showed signs they were nursing pups. Soon after that I saw all twelve Unknown wolves camped out in the Druids' Chalcedony rendezvous site, which was about seven miles west of the new Druid den. The Unknown alpha pair did a lot of scent marking. This could be a sign that the pack, now that they had the Slough wolves on the run, were planning on taking over the Druid territory.

I noticed that the Unknown wolves' denning female was in the group. The pack was back at Slough Creek the next morning but the female did not go into her den, which indicated that none of her pups had survived. That could be a game changer for the Slough wolves because the Unknowns were no longer tied to the area. It also meant the invading wolves could now lay siege to the new Druid den if they discovered its location.

A few days later, the Unknowns howled from close to the old Druid den forest. Later that day, the four Druids were feeding on a carcass just a mile from the Unknown wolves, south of the road. In late evening, the Unknown wolves howled to the north. 302 was the first of the Druids to react. He ran away from the carcass and the others followed. 480 positioned himself behind the two females. I had seen former Druid alpha male 21 do the same thing when his family was threatened by rival packs. That custom of alpha males placing themselves between their pack and a perceived danger demonstrated the depth of their commitment to their families.

Then we heard additional wolves howling to the west. It was the Sloughs. The Druids were close to two rival groups of wolves, a very dangerous situation, especially because their new den was only a few miles away. The Druids did not howl back at the other wolves. That silence saved them, for neither the Unknowns nor the Sloughs knew they were there. As they did not have the numbers to stand up against the other packs, the Druids had to avoid getting in battles until their pack was larger. By the next morning, both the Sloughs and Unknowns had left the area.

We saw the four Druid adults on a fresh carcass at Round Prairie early on May 23. The pack was making frequent kills, so the two mothers were probably producing plenty of milk for their pups. 302 was by himself in that meadow later that day, but he was not feeding at the carcass. Then I noticed that several people had walked toward the site. I went out and explained that their presence was probably keeping 302 from eating, and they all agreed to help him by going back to the road.

In early June, I saw 302 carrying the head of an elk calf through Round Prairie. Mother wolf 569 grabbed at the head as he went by and the two wolves ended up in a tug-of-war. 302 won, and I watched as he disappeared in the direction of the den in the trees still holding on to his prize. I guessed he intended to give it to the pups rather than eat it himself.

After 302 left the meadow, the other three adults chased and killed a cow elk. 529 took out the liver and ran off with the delicacy. The liver and other internal organs have more nutritional value to wolves than red meat, so those parts of a kill are of great value to a nursing mother. Later she and 480 each carried an elk leg back to the pups at the den.

The pack made another kill in Round Prairie three days later. I was gradually realizing what a good choice the females had made when they decided to den near here. Not only was the den site in the trees far away from rival packs, but the Druids were also having a lot of successful hunts in the meadow at the exact time the family had a lot of extra mouths to feed. The single most important decision an alpha female makes over the course of the year is where to den. 529 was just two years old and had never had pups before, but despite her inexperience she was proving to be a superior leader. I felt with her as alpha female and 480 as alpha male, the Druids could have a dramatic resurgence.

ON JUNE 11, I completed my sixth year of going out to study wolves every day. That day Ed Bangs, the U.S. Fish and Wildlife biologist who was the leader of the Wolf Recovery Program for the Northern Rocky Mountain states, joined me in Round Prairie to watch the Druid wolves.

To keep the public informed, Ed and his co-workers had traveled around the region prior to the reintroduction of wolves into Yellowstone and Idaho and held a large number of public meetings on the issue. The meetings were often very contentious because many of the attendees did not want wolves brought back to the region. Ed has an easygoing personality that was perfect for those meetings. Once a local man angrily demanded that Ed tell him one thing wolves were good for. Ed answered with a line that got a big laugh from the audience: "They gave me this cushy government job." Wolves might never have been reintroduced in Yellowstone if some officious government bureaucrat had been in Ed's job.

A lot of people had gathered around us at Round Prairie, and as I helped them see the Druids through my scope, I filled everyone in on the story of each of the wolves we could see. As always happened, they especially loved hearing about 302. Then I pointed out Ed and explained how he had done so much to help get wolves back in the park, and everyone spontaneously gave him a round of applause. Ed is a very modest man who seemed a little embarrassed at the praise but he deserved it.

WE SPOTTED FOUR Unknown wolves howling at the Druids' Chalcedony rendezvous site in mid-June. They apparently were away from the rest of their pack. A few minutes later, the four Druid adults came into that area, probably to find out what four wolves were in their territory. They spotted the other wolves and charged at them.

The Unknowns jumped up and ran off. The Druids pursued them with 480 in lead. He reached a yearling black and pinned him. The other three Druids ran in and all four bit the yearling a few times. They let the young wolf go and chased other wolves. All eight wolves met up at the edge of the trees and fought until the Unknowns lost their nerve and fled. The Druids chased them and bit at their hindquarters. 480 led his pack as they routed the Unknown wolves.

That turned out to be the final time I saw the Unknown wolves. The irony was that the last of the Unknowns were driven out of the area by the Druids, under the leadership of 480, rather than by the much larger Slough pack. Unfortunately for the Sloughs, they were to have one more run-in with these wolves, and it would not end well for them.

After the incident with the Druids, I thought a lot more about leadership. 480 saw what needed to be done and did it without hesitation. He set an example with his actions, and the other Druid wolves backed him up and followed him into battle.

That reminded me of an example of leadership in a dangerous situation from my college years. President Nixon expanded the Vietnam War in the spring of 1970 by bombing Cambodia and sending troops into that country. At that time, American soldiers in Vietnam were dying in combat at the rate of 428 per month. Students at seven hundred colleges across the country protested by going on strike from classes and holding mass demonstrations against the war. Four students at Kent State in Ohio were shot and killed by National Guard troops and another two were killed at a college in Mississippi by state police.

There was a violent segment of the anti-war movement that in the previous year had set off 120 bombs in campus buildings across the country. During the strike, thirty colleges had bombings. Later a bomb set off in a Wisconsin university building killed a student and injured three others.

At my school, the University of Massachusetts, there was a mass meeting of students, and the violent faction announced plans to place a bomb in Memorial Hall, a prominent symbol of the university. Our side, the nonviolent contingent, tried to talk them out of the plan but made no headway.

Then a student stood up and said if they were going ahead with their plan, he would camp out in that building. If a bomb went off, it would kill him. The other side still would not relent. Several of us backed up our guy and I signed up for the night shift at Memorial Hall. I would eat dinner at the cafeteria, bring my books and a sleeping bag to the building, spend the night, then leave in the morning when I was relieved. We did that for a week, and it saved the building.

That was a crisis where an individual stepped up and volunteered to put himself at great risk. His actions caused others to follow his example and it kept the peace. I equate that story to how wolves like 8, 21, and 480 would see a rival pack threatening their family and instantly charge forward, regardless of the danger to themselves. Their decisive actions caused the other wolves in their pack to follow them into harm's way. That aspect of leadership, in wolves and people, is exemplified by the motto of the U.S. Army Infantry regarding leadership: Follow Me.

But leadership in people and wolves more often involves issues that are not life or death—it is taking care of details and responsibilities for everyday matters rather than ignoring them. When I worked at Death Valley National Park, I was often the supervising ranger when movies were filmed in the park. I had the authority to shut down a shoot if there were any problems with the crews.

I was on location when one of the biggest films of all time was made in Death Valley. Everything was going well and there had been no issues to deal with. Toward the end of the shoot, I saw that a crew member was smoking a cigarette and thought he might throw it on the ground when finished. It would be a minor offense, and I figured I would just go over and pick it up and not make a big deal about it.

He did throw it on the desert floor. Before I could start to head that way, another person on the crew quietly walked over, picked the butt up, and put it in his pocket without saying a word. I looked around and noticed that everyone else had seen what he did.

That was George Lucas and the movie was *Return of the Jedi*. He saw what needed to be done and did it. I had sensed that George had great leadership skills and that his crew would follow his example in everything he did. Picking up that cigarette was a small action that took only a moment, but it made a big impression on his crew. Rather than telling them what to do, he showed them what to do. The college student and George Lucas both led by example, the same type of leadership style I had seen in the great alpha wolves.

I SAW AGATE alpha female 472 with her new pups at Antelope Creek, south of Tower Junction, in late June. There were six pups: four blacks and two grays. One gray pup would grow up to be the 06 Female and the other would be known as 693, who was destined to play a big role in 302's life. All six pups slept together in a big pile, looking like a collection of children's stuffed animals.

Later that day, I spotted the Druid alphas 480 and 529 in Round Prairie with five pups following them. 302 seemed to be herding the pups from behind. When the alphas got well ahead of the pups, the pups sniffed at their parents' trail, showing that they already knew how to follow a scent even though they were barely two months old. 302 looked around a lot, like he was scouting for additional pups. A few days later I counted eleven pups: seven blacks and four grays. They played in the meadow and fed on an old carcass. At one point, 529 howled and the pups howled back. Seeing all those pups reassured me that the Druids were in a renaissance.

I saw 302 babysitting the pups in the meadow the next day. After bedding down for a while, he got up and the pups ran to him. When 302 went into the trees, the pups dutifully followed him. Another day I saw him regurgitate meat for the pups. He might not have been cut out to be the Druid alpha male, but he was doing a fine job running the pack's day care center.

12

Summer

ON JULY 2, wolf 302 was once again taking care of the eleven pups while the other adults were elsewhere, probably on a hunt. He must have just visited a carcass, for he gave the pups lots of regurgitations. The pups surrounded him and wagged their tails, repeatedly licking his face to get him to regurgitate more meat.

The next morning, the Druid adults were nowhere in sight, but the pups were in Round Prairie entertaining themselves by chasing, wrestling, and ambushing each other. There were several recent carcasses nearby for them to feed on. They would occasionally howl, trying to contact the older wolves. The adults were back with the pups on the evening of July 4.

Early the following morning, the only Druid in the meadow was a single black pup. It was doing a lot of howling and looking around for other pack members. I later learned the pack had moved the other ten pups out of the meadow and that the black pup had apparently refused to follow. The next day the pup continued to howl. Mosquitoes and biting flies were pestering me that day. They must have been bothering the pup as well for it repeatedly snapped in the air. Still snapping at mosquitoes, the pup went to an old carcass and fed, then carried off a bone and bedded down with it. Meanwhile Dan Stahler, who was out on a tracking flight, found the Druids up Cache Creek, about six miles from the pup.

On July 9, the four Druid adults showed up in Round Prairie. The pup had been alone for five days. Young female 569 and the pup met up and romped around together, then alpha female 529 ran in and joined in the play. When the pup saw 302, it ran over to greet him, then returned to 569. She spent the most time with the pup, which probably meant she was its mother. I lost the wolves in the trees and the signals from the adults soon faded away. They were likely taking the pup back to the rest of the family. Emily Almberg did the next tracking flight and got the Druid signals at Cache Creek, and hikers reported seeing pups in that area.

Later in the season, after the pack had left, a Wolf Project crew investigated the Druid den east of Round Prairie. It was under a rock overhang and looked like a cave. There was mountain lion fur there, indicating that a female might have used the site as a den for her kittens before the wolves moved in.

The Slough wolves mostly stayed out of the Druid territory during that time, but the alphas and female 527 came into Lamar Valley in mid-July. When they reached the Chalcedony rendezvous site, they sniffed around where the Druids had fought and defeated the Unknown wolves a month earlier. The Sloughs followed scent trails around the area for an hour. Wolves, like bloodhounds, have a finely tuned sense of smell, which means they could probably separate the scent of the Druids from the Unknown wolves and figure out that the Druids had chased off the other pack.

I later learned more about wolves and their sense of smell from local naturalist and wildlife researcher Jim Halfpenny. One winter day, he had watched as a traveling wolf stopped and stuck its nose into the snow to sniff at something. When he dug out the site after the wolf had left, he found a wolf scent mark twelve inches below the surface. The wolf had gotten the scent through that much snow as it was walking by.

THERE WAS SNOW on the ground when I left my cabin early on September 17. I had long underwear on that day and was wearing my winter boots despite the fact that technically it was still summer.

In late September, ten Agates chased the Slough wolves when the smaller pack ventured into Agate territory south of Slough Creek. I had known Agate alpha male 113 since he was a yearling. He was nine years old now, the same age 21 had been when he died. Despite his advanced age, 113 was still serving his pack well, driving out a rival pack that had entered his territory. Like 480 he had good leadership skills and acted decisively when necessary to protect his family.

Slough alpha male 490 did not stand up to the elderly 113, just as he had not stood up to the Unknown wolves when they invaded his territory. His failure of leadership led to the death of all the pack's pups and the killing of his two brothers: 377 and 489. I thought that 302's pattern of running away from fights was unique to him, but now saw that 490 seemed to have a similar character.

In mid-October, I saw fifteen Druids coming into Lamar Valley from Cache Creek. All eleven pups had survived their first six months, a period when pups are often lost to malnutrition, disease, and conflicts with rival packs. That was a testament to the high quality of care by the four adult wolves. The family seemed to have been on a short reconnaissance mission, for they soon turned around and went back to Cache Creek. There had been no rival wolves they needed to drive out of their territory.

A few days after that, the Sloughs returned to Lamar Valley. I lost sight of them heading toward Cache Creek. Thirty-eight minutes later, I spotted 490 back where I had first seen his group. He nervously looked around and seemed to be searching for the rest of his pack. 490 howled and female 526 ran to him. They raced west, stopped repeatedly to look back in what seemed to be a fearful manner, ran west again, and were joined by other members of the pack. 490 headed up into the trees to the south of the valley and the others followed.

I checked for the Druid signals and got them toward Cache Creek. Now everything fit together. The Slough wolves must have run into the Druids. Outnumbered two to one, the Sloughs had fled. The Druids won that encounter. It was especially fitting that the incident took place at Cache Creek, where the Slough wolves had killed Druid yearling female 375 the previous year.

In early November, the Slough alphas and 526 traveled to the Chalcedony rendezvous site and one of them howled.

The fifteen Druids were across the river less than a mile away. This time they must have known they outnumbered the Sloughs and howled back. The intruding wolves looked in the direction of their howls, then retreated.

The Druids soon arrived in the rendezvous site and sniffed around. 480 was out in front of the two adult females and the eleven pups. Last in line was 302. I wondered if he had positioned himself there to make sure none of the pups wandered off or if he was wary of running into rival wolves.

A few days later, the Sloughs killed a bull elk in the rendezvous site but abandoned it when the Druids came in from the east and claimed it as their own. Later in the month, there was a howling contest between the two packs and the Sloughs once again retreated from the Druids. Those incidents proved that the Druids had come back from the brink and were once again the dominant pack in the region.

In early December, we spotted the Druid wolves near the Lamar River. Three black pups were missing and alpha female 529 was not in the group. Her sister, 569, did double scent marks with 480. When I found the pack two days later, 529 was still absent. 302 was bedded down, and 569 went to him wagging her tail. She licked his face, then jumped around him in a flirty manner. The mating season was coming, and she was showing her interest. 480 noticed, came over, and pinned 302.

BY THE MIDDLE of December, there was deep snow in many areas of the park. It got down to minus 20 degrees Fahrenheit (–29 Celsius) one morning that week. Around that time, I saw some of the Druid pups playing on a snowy slope. One of them dug into the snow, causing pieces of crusted snow to glide downhill toward another pup. It studied the chunks and tried to grab them as they slid by.

I last saw Slough alpha male 490 with his family on December 19. After that we spotted the other seven Sloughs several times, but not him. Then Doug got a mortality signal from 490 way up Slough Creek. He searched for the wolf from the air and spotted his body on the ice in the creek. A crew went up there the next day. 490 had been dead for four or five days and evidence at the site showed he was killed by other wolves. That area was a place where the Unknown wolves were still ranging, so they likely killed him, just as we suspected that they had killed 377 and 489.

His loss meant that now all the wolves in the Slough pack were females. The mating season would be starting in a few weeks, and the pack needed a replacement alpha male right away. I immediately thought of 302. He might like to be the only male in a pack where he had a harem of seven females. I felt he was becoming more mature and responsible. Perhaps this was his big chance to be an alpha. But another male acted faster and beat him out: a gray Agate yearling. With all the drama and trauma the Sloughs had been going through, the young male was going to face a lot of difficult challenges. Maybe 302 was better off not trying to join the pack.

That day the new Druid alpha female, the one I have been calling 569, was darted and radio-collared, along with two gray pups: male 570 and female 571. The missing three black pups never returned and must have died. We never saw 529 again and had to assume she had also died. Since there were eight healthy pups and three collared adults—480, 302, and

569—the Druids now had a good chance to regain their dominance of Lamar Valley. 302 seemed destined to live out the final years of his life in a supporting role as the pack's second-ranking male.

PART 4

2007

13

302's
Walkabout

I SAW THE AGATE yearling gray male with the Sloughs on the second day of the year. The pack fed on a new bison carcass, then all seven females went to the male and flirted with him. He seemed confused about getting all that attention. He snapped at one female, then at two others. His hormones had probably not kicked in yet for the approaching mating season, his first one. The females clustered around him, pressed against his sides, jumped on his back, and licked his face. It looked like old film footage of teenage girls mobbing Elvis or Paul McCartney.

A coyote approached the carcass and the gray male charged after it. He grabbed its tail and pulled the coyote down. It turned and tried to bite the wolf, but the yearling got a holding bite on the back of the coyote's neck and shook it. The females all ran in and killed it. That incident showed that, at least when it came to defending his family, the pack's new alpha male was capable and quick to act despite his young age.

That day the young alpha male did a lot of scent marking and four of the Slough females marked over his sites. Flirting and marking sites were signs of his full acceptance into the pack. He bedded down and tried to get some rest, but a young female came right over and pawed at his back to get his attention. He snapped at her. Another female arrived and did the same thing to him. He was not going to get much rest with seven females vying for his attention.

THE DRUIDS NOW numbered eleven: three adults and eight rapidly growing pups. Although the Sloughs numbered only eight, seven females and one male, they potentially had an advantage because all eight Sloughs were adults.

It was minus 35 degrees Fahrenheit (–37 Celsius) when I left my cabin early on the morning of January 12. The low out in Lamar was minus 40 and it got to the same temperature the next day as well. It was warmer the following morning: minus 30 degrees Fahrenheit (–34 Celsius). This was my eighth winter in Silver Gate, but I still had not acclimated to the cold. On days like that, I often thought of the warm winter weather I had experienced when I worked in Death Valley National Park.

A few days later, the Druids and Sloughs were both on the north side of Lamar, a few miles apart. The two packs howled at each other. Later the Sloughs moved off to the west and the Druids went east. It looked like the leaders of both packs had chosen to avoid a confrontation.

302 was not with the rest of the Druids on the afternoon of January 16. He was seen the next day at Slough Creek, following a scent trail, probably of some of the Slough females. By evening he was back with his pack. It was too early for females to breed so 302 must have been on one of his reconnaissance missions, checking out available females in the area. Meanwhile the new Slough alpha male was getting more comfortable with all the female attention. He now often led the pack when they traveled. For a young male, he seemed confident.

I saw 302 following a scent trail in the Slough Creek area later that month. Once again, he apparently failed to find any females for he was back with the Druids the next day. 480 dominated and pinned 302 more than usual, likely in anticipation of the upcoming mating season. Soon after that, 302 tried to interact with new alpha female 569, but 480 rushed over and got between them. Later Laurie Lyman twice saw 302 get up in a mounting position on 569. Each time 480 saw them and came right over, and 302 dropped off.

After that Laurie saw an example of 480's watchfulness over his family. When two pups approached a golden eagle on the ground, 480 came over and got the pups away from the bird. Those pups later returned to the eagle and 480 once again drew them back. I had once seen an eagle grab a pup's chest with its talons. The pup got away but could have been seriously injured. The talons of an eagle are its weapon, rather than the beak, and the big birds have enough gripping power to kill small animals the size of wolf pups.

By the second half of January, 302 had a pattern of staying with the Druids for a while, then taking off and slipping into the Slough pack's territory. He would bed down near them and howl back when they howled. That communicated to the females that he was nearby and available for mating, the wolf version of Tinder.

John and Mary Theberge, two wolf researchers from Canada, were in the park doing a long-term study of wolf howling. They later analyzed ten years of my field notes and found that wolves howled most during the February mating season and least during the spring denning season. The low level of howling at den sites was likely because packs were being secretive about the location of their pups. Much of the February howling was probably young males and females calling out to prospective mates.

302 went on another scouting trip into the Slough pack's territory at the end of January. Three days later, I got his signal and signals from Slough females from the same area. After a few minutes, I spotted them. Slough alpha female 380 was approaching 302 and teasingly jumping around him, showing her interest. But 302 looked past her and tucked his tail. I scanned that way and saw the pack's alpha male. 302 ran off from the much younger and smaller wolf. He still seemed to have a fear of getting into a fight. 302 failed to breed that day, but I later had reason to think he got away with at least one mating with a Slough female.

As I checked on other packs that January, I realized that Agate alpha male 113 had been apart from his family for nine days. I found him alone and watched as he got up from a bedding site and limped on one of his hind legs. After just a few steps he lay down again. It looked like it was painful for him to walk. I spotted him again a week later. He had lost weight and was still alone.

I saw him with the Agates the next day. He was mostly last in line as they traveled, and I saw him lick his hindquarters a lot. Alpha female 472 came over and greeted him. 113's son, beta male 383, joined them, his tail wagging higher than his father's, which hinted that the pack hierarchy was changing. The pack moved off and 113 followed at a slow pace. The Agates bedded down and waited for him. When 113 caught up, he immediately lay down by the others.

On the nineteenth, I got a better view of 113 and saw a bloody area on his rear end. That must have been the spot he had been licking. The flesh there was torn, indicating the wound was not from an elk kick. He might have gotten it in a fight with a mountain lion. Although he managed to keep up with the pack most of the time, 113 never fully recovered from that injury and there was a peaceful transfer of the alpha male position to 383. From that time on, 113 lived in sort of a retirement phase of his life.

I had always admired 113. He was the founding alpha male of the Agate pack, and his leadership style was to treat the other males in the group fairly and nonaggressively. I thought of him as a team player, and his amiable relationship with 383 and the younger Agates was greatly benefiting him now that he was dependent on them in his old age.

I soon witnessed a touching moment involving 472, his longtime mate. She went to 113 as he was lying down and stood beside him with her tail averted, her way of indicating she wanted to mate with him. He struggled to get up, then dutifully sniffed her rear end. With his injury, he could not do much more than that, so he lay down again. Another day, two pups accidently bumped into him while they were playing, something I thought would anger an old wolf who was probably in a lot of pain. I expected him to snap at them, but 113 just wagged his tail at the pups. After that I saw 472 bed beside 113, a gesture that must have been comforting to him.

A JANUARY TRACKING flight circled over the Mollie's wolves, and Doug noticed that 193, their old alpha male, had bare patches on his coat. Later that month, Doug collared another adult male from that pack who also had missing fur. That was very concerning for it indicated they had mange, the first two documented cases in park wolves.

Mange is caused by infestations of parasitic mites just below the skin. Rebecca Raymond, a biologist who worked for the Wolf Project, had previously been a veterinary technician. She told me the mites live at the base of hair follicles. When a dog or wolf scratches at the spot, it pulls out fur along with the mites. The loose mites can then easily get on another wolf in the group and the infestation spreads much like lice in children.

Historical records document that mange was deliberately spread among wolves and coyotes in the American West as part of a bigger effort to eradicate those predators. It was a form of biological warfare. In 1905 the Montana legislature passed a bill instructing the state veterinarian to oversee a program to capture wild wolves, infect them with mites, then release them "in hopes that they would return home and infect their fellow pack members." The bill was entitled "An Act to provide for the extermination of wolves and coyotes by inoculating the same with mange."

I consulted with Ellen Brandell, a PhD student at Pennsylvania State University who was doing research on wolf diseases and parasites in our area. She told me that when the early Yellowstone rangers killed off the last of the park wolves in 1926, mange continued to reside in the regional coyote population. This was the source of the infection we were now seeing in the park's wolves.

Since we were approaching the mating season, the timing was especially bad. A male from the Mollie's wolves with mange could meet a female from another pack and spread the mites to her. If she later returned to her family, the mites might get on them. Mange does not kill a wolf outright, but the infestation weakens the animal and the loss of fur greatly reduces the afflicted wolf's ability to cope with extremely cold weather. A wolf suffering from mange has a much better chance of full recovery if it is a member of a pack that will feed it and protect it from rival wolves. Lone wolves have a harder time surviving. Unfortunately, a wolf that recovers from mange can become reinfected if it is exposed to new mites.

BY EARLY FEBRUARY, the breeding season for the wolves was well underway. On the third day of the month, the Druid alphas, 480 and 569, mated. They got in a second tie nine and a half hours later. A few days later, they had a third mating.

On February 4, the Agates' new alpha male, 113's son 383, mated with alpha female 472. Dan Stahler told me that genetic analysis showed that 472 was his aunt. 383 also bred 471, a younger female, who was his half sister. Occasional matings over generations between related wolves in a pack probably do not cause much harm genetically, but if that happens more often there will be serious inbreeding problems.

Rolf Peterson, the biologist who has studied wolves at Isle Royale National Park for decades, told me about the problems caused by inbreeding there. Wolves colonized the island in the late 1940s by walking over the frozen Lake Superior. Within fifteen years of the wolves' arrival, skeletons of island wolves showed abnormalities in their spinal columns, evidence of genetic problems caused by inbreeding. Rolf told me that all wolf skeletons they examine these days have deformities. The much larger Yellowstone wolf population does not have that problem because young wolves are usually able to find unrelated mates.

Two new males came into the territory of the Slough pack. First a black outsider mated with subordinate female 526, then 590 from the Agate pack arrived. He seemed to be a brother of the Sloughs' new alpha male because the two had a friendly greeting. Soon after that, the young Slough alpha male mated twice with alpha female 380, then mated with 527, the pack's beta female.

While all this mating activity was going on, I did not pick up 302's signal. I thought he was probably avoiding the males hanging out with the Slough females and trying his luck elsewhere. I finally got signals from him to the north from Slough Creek on February 10. There were no Slough signals that way, which suggested he was with some of the Unknown females. This was a risky move since the Unknown wolves had killed three of the Slough males.

In mid-February, I spotted a mother lion with three kittens, so I took a break from watching wolves and put my scope on them instead. The female was up in a tree near a fresh elk carcass. She came down and two of the kittens played by batting their mother's long tail back and forth. Later one of them put a front paw around the female's head and they wrestled. Another kitten came over and licked her face and she licked it back.

I got a strong signal from 302 on February 17 and saw him alone near the Druid den forest. He was limping badly on his hind left leg. I spotted a wound there and figured he had been bitten by a rival male competing with him for a female. After a while, 302 howled and then moved off, probably to search for his packmates. The snow was deep and 302 tried to hold his injured leg off the ground as he walked, but he frequently fell through the surface crust and had to use that leg to pull himself out.

302 was heading toward the road, intending to cross to the south. Two cars stopped, blocking his route. He turned around and struggled to go back through the deep snow with his bad leg. We asked the drivers to help by moving on, and 302 got across once they left. He then went east, looking for the Druids.

He stopped off at the carcass of an elk the Druids had killed three weeks earlier. All that was left was bones, so he did not get much of a meal. He howled in another attempt to contact his family, but there was no answer. The next day, he searched for the Druids in Round Prairie. He was farther east on February 19. That was a mistake, for I found the other Druids in the opposite direction with a fresh elk kill. On February 26, I got his signal and the other Druid signals from the trees in Round Prairie, indicating he was finally back with them.

The reunion, however, was brief. A few days later, 302 was once more apart from the other Druids, probably because his injured leg made it difficult for him to keep up with the pack. We had noticed that there were other wounds on his rear end and right hip. 302 probably had been beaten up several times during his romantic adventures with the Sloughs and the Unknown wolves. It was a price he had to pay for wandering around in the territories of rival packs by himself, trying to mate with their females.

When I saw 302 in early March, his hind leg was still stiff as he limped along. When that paw was about to touch the surface of the snow, it twisted to the left at a forty-five-degree angle. Jim Halfpenny, who is an expert on animal tracking, later looked at 302's trail in the snow and told me that twist indicated there was a broken bone in the leg. 302 often held that paw off the ground as he traveled, trying to avoid the pain of putting weight on it.

Mark Rickman is a Colorado physician who, along with his wife, Carol, spends a lot of time studying wolves in the park. He told me what might happen to 302's leg. New bone material in both humans and wolves is generated by cells called osteoclasts. Those cells, along with calcium, mend a fracture by forming a deposit called callus. I hoped that natural process would heal 302's leg over time.

Several days later, I heard the main Druid group howling. 302 howled back. Wolf Project biologist Matt Metz flew over that area and called down to say he saw 302 approaching the Druids in a submissive crouch. All the wolves ran to him and excitedly mobbed and greeted 302 for several minutes, welcoming him back to the family. The next day, when the pack traveled away, 302 stayed behind—probably because his injured leg was still making it difficult for him to keep up. He later joined them at a carcass site.

302 HAD A complicated history with carcasses. Back in March 2006, wolf watcher Kathie Lynch saw the Druids chase and attack a bull elk. 302 must have been kicked during the battle for he lay down after they killed the bull and did not move for some time. When he finally got up, he was limping. 302 approached the carcass and fed briefly, but he seemed wary of the dead elk and repeatedly backed off. When the carcass moved as another wolf pulled off a strip of meat, 302 stepped away like he thought the elk was still alive. He later returned to the carcass but once again jumped back from it as though he was afraid. I had seen young pups do that at a carcass, but never a big adult male. After that he avoided the carcass and scavenged on scraps of meat other wolves had carried off from the site.

302 acted the same way a few days after that when the Druids killed a young bull elk, jumping back several times while he was feeding. He seemed to be thinking the bull was going to attack him. I thought back to a time when I was watching 21 feeding at a bull elk carcass. When another wolf yanked at the carcass, one of the elk's sharp antler points poked 21 in the face. 21 paused briefly, then resumed eating.

In mid-March, I saw the Druids at a new bull carcass. 302 was resting away from the elk. When a pup left the site after feeding, 302 jumped up, went to the pup, and licked its mouth, trying to get it to regurgitate meat. The pup lowered its face and brought up a pile of fresh elk meat. Both 302 and the pup fed on it. It looked to me like 302 was too wary of the bull to feed on the carcass. If that was indeed the case, his solution was to wait for the pup to feed, then get it to regurgitate some of the meat to him.

I watched as 302 approached the carcass on five separate occasions only to back off each time. He just did not seem willing to feed on it directly. When he next went toward the carcass, he tried to steal a piece of meat that a pup had just pulled from the elk, but the pup stubbornly held on to it. Then 302 tried, unsuccessfully, to snatch meat from another pup. 302 went to the carcass a seventh time, but when he got close to it, he jumped away in fear. Later 480 regurgitated meat for the pregnant 569 and a pup. 302 ran over, butted in, and got some of their meat. Still afraid of the dead bull elk, 302 walked around the other wolves, looking for scraps they might have dropped. 302 was regressing. Last year he was regurgitating meat for pups. Now he was stealing food from pups and a pregnant female.

Soon after that, when 302 was apart from his pack, he spotted a group of wolves and ran away. 302 stopped twice, looked back, and ran off each time. I put my scope on the other wolves and saw that it was the Druids. There were Mollie's wolves in the area. 302 must have gotten their scent and assumed these wolves were from the Mollie's pack. But it was his own family. He crossed the road and hid out in the old Druid den forest north of the Footbridge lot. The incident suggested that wolves' long-distance vision is not good enough for them to identify other wolves if they are far away. Unable to tell if the wolves he had spotted were his family or a rival pack, 302, staying true to his personality, ran away.

As I thought about 302, I wondered if he had gone through some type of traumatic event when young, something that was now shaping his behavior as an adult. What triggered that idea was my experience with a friend's dog, a big German shepherd named Riker. He was a rescue animal from a shelter who mostly behaved normally but exhibited extreme fear in certain situations. My friend lived in Cooke City, four miles from my cabin. I sometimes stopped in while wearing my park ranger uniform and badge. As soon as Riker saw me, he would cower and try to hide. But if I arrived out of uniform and without the badge, he was always friendly. We figured that he had been abused by a man wearing a uniform and badge. I made a special effort to make friends with the dog and eventually Riker behaved normally even if I showed up in uniform. That made me wonder if a big male had come into the Leopold pack when 302 was a pup or yearling and apart from his father and the other adults. If that wolf had attacked and severely beaten up 302, I could see how it might have traumatized him for life and made him fearful of being killed in a fight with other wolves.

14

The Battle of Mount Norris

THE BIGGEST EVENT of 2007 took place on March 22 when 302 was away from the Druids. He was resting near an old moose carcass, south of the road, near the northern base of Mount Norris. I spotted Mollie's wolves up high on Norris and got a count of ten at or near a bison carcass: five adults and five pups. Their old alpha male who had mange was not with them, which likely meant he had died. Our March Winter Study crew, Abby Nelson and Scott Laursen, found the rest of the Druids near Chalcedony Creek, about a mile to the west. There were nine of them: 480, alpha female 569, and seven pups.

The Mollie's wolves howled. The Druids howled back, looked toward Norris, and did more howling. Within a few minutes, 480 started marching toward the Mollie's wolves followed by the other Druids. He led his family across the

Lamar River and up the western slope of Norris. He was doing exactly what 21 would have done in such a situation, advancing directly at a group of wolves that had invaded his territory. But in this situation, his strategy of confronting the opposing pack was risky.

There was only one other adult on his side, the alpha female, and she was pregnant. 569 was too important to the pack to risk putting in a battle situation. Plus her pregnancy was advanced enough that it would hinder her ability to run, either at the rival wolves or from them. The Druid pups had never been in a battle before and probably would run off if 480 started a fight. If it came to a confrontation, he likely was going to be the only combatant on his side. It would be him against five adult wolves.

The Druids reached the bison carcass and sniffed around. 480's hackles were up as he got the scent of the other wolves. Then he followed their scent trail uphill. Six pups followed him. The seventh pup and 569 stayed behind. I lost sight of the Druids as they crested a ridge. That was where we had last seen the Mollie's wolves. Around that time 569 slipped away, which was the right thing for her to do.

A few minutes later a lot of wolves appeared, running in different directions. Four Mollie's wolves pulled down a Druid pup and attacked it. Then 480 charged in with five pups following him. He chased, knocked down, and attacked a Mollie's wolf, then let it go. As that was happening, the Druid pup that was being bitten escaped and ran off.

A black male charged at 480, who instantly counter-charged, causing the black to run away. Seeing the rival wolf flee, 480 and his pups regrouped and stood defiantly in the middle of the battlefield, looking like they were eager to take on any other challengers. 480 chased an approaching Mollie's wolf, caught up with it, and pulled it down. He had the animal at his mercy but chose to let it go. As it ran off with its tail tucked, I saw it was a pup. Scott watched 480 smash into the side of a gray and knock it down. Rather than attack the downed wolf, 480 allowed it to get up and flee.

480's strategy of taking the initiative and charging at the opposing pack, despite having only pups on his side, was working brilliantly. As I looked around the battlefield, I saw Druid pups chasing a big Mollie's male who had his tail tucked, a sure sign he was intimidated by his much smaller pursuers. Elsewhere 480 knocked another gray wolf down, then chased the gray when it jumped up and ran off.

Then 480 spotted the Mollie's alpha female, who was with other pack members, and charged at her. Her group ran away and 480 let her go. Druid pups then gathered around their alpha male. Two of the big Mollie's males ran at them. Without hesitation, 480 countercharged them. They lost their nerve and fled. In the next few minutes that scenario was repeated several times: Mollies' wolves would reorganize, charge at 480, then run off when they saw him coming at them.

The pups reunited with 480 and things calmed down. He walked over to a snowbank and bedded down on it, probably trying to cool off after all that chasing. Laurie Lyman was with me and noticed that the four pups who had stuck with 480 during the most intense moments of the conflict were all female. One was 571, who years later would display extraordinary courage when she successfully protected a group of Druid pups from a rival pack. She was a daughter of 480 and granddaughter of 21 and had clearly inherited their battlefield courage.

The biggest of the Mollie's males then headed downhill toward 480. The Druid male howled and the four pups with him joined in. As 480 stood his ground protectively next to those pups, the other big male turned around and retreated. Soon after that, now that 480 had won the conflict, alpha female 569 and the other pups rejoined him.

Throughout the battle, when 480 was so decisively defeating the Mollie's wolves, 302 was comfortably resting by the moose carcass, about a mile away. He must have heard the nearby Mollie's wolves howling, the howls that caused 480 to march directly at them. 302 chose to ignore the calls of the invading pack and spent the day safely feeding and napping, well away from the battlefield. I had been rooting for 302 to do better in life for over four years, but his total noninvolvement as 480 rushed to confront these wolves made me think 302 was a lost cause. It seemed like he had given up on even trying.

WHEN I LATER thought over the confrontation, I was struck by one aspect of 480's behavior: he never attacked a rival wolf all out. He would catch up with one, knock it down, bite it a few times, let it go, then look around for other Mollie's wolves to go after. The four female pups loyally stayed with him and helped 480 chase and intimidate the other wolves into running away, one after another.

I realized that 480's strategy of catching and briefly biting at a downed wolf, then going after the next one, played to the strength of the pups. To them this probably seemed like a chasing game, the kind they would play among themselves when they would pick one pup to chase, tackle and nip at it, then go after another littermate. The pups were now nearly eleven months old, and it was probably hard for the Mollie's wolves to tell them from adults at a distance. When they saw 480 and four wolves charging at them, they likely thought all were adults and fled. As he commanded the battle, 480 used the troops that he had, four inexperienced female pups, about the same age as fourth-grade girls, and charged forward, intimidating a group of larger-than-average adult wolves into running off without putting up much of a fight.

I later did a wolf talk for a group of visitors and told them about the battle of Mount Norris. After I finished, one of the men in the crowd came up to me. He was a retired soldier who had been in a lot of combat. The man told me that 480 used a classic military strategy that was described in the ancient Chinese book *The Art of War*: if you are weak, pretend to be strong. It means if the enemy outnumbers your side, take the initiative and charge at them. They may falsely assume that your army must be superior to theirs, panic, and run away. That is exactly what 480 did. The veteran mentioned that he occasionally lectured at the Army War College and suggested that he and I do a joint talk where I would describe the strategy, tactics, and leadership methods that alpha males like 8, 21, and 480 employed in battle situations. I wondered if I should contrast that with 302's preferred strategy of avoiding combat altogether.

15

The Druids and the Sloughs

OFFICIALLY IT WAS spring, but in early April we got twelve inches of snow at our high-elevation town of Silver Gate. By that time, a number of female wolves in Yellowstone were in advanced stages of pregnancy, including Druid alpha female 569 and six of the seven Slough females. There was still just one male permanently with the Sloughs, the young gray from the Agate pack. We had seen 302 near the Slough wolves during the mating season, and I guessed that he had gotten some of those six females pregnant.

I saw the new Slough alpha male lead a chase after a bull elk who had shed his antlers. The elk must have been in poor health for the wolves easily caught up with him. The bull tried to defend himself by kicking back at wolves behind him and by lunging his head at the ones in front of him. Since his antlers were gone, his lunges were harmless. The wolves

soon finished him off. The next morning, a grizzly that probably had just come out of its winter den took that kill away
from the Slough wolves. They waited for it to finish, then fed.

The following day I saw grizzly tracks in fresh snow near
my cabin. We later found out that it had broken into a barn
across the street, torn open a bag of birdseed, and eaten
some of it. After that, park ranger John Kerr, who is a summer
resident in a cabin near my place, had a grizzly on his front
porch. Another time I found bear paw prints on my car when
I went out to start it in the early morning.

By mid-April, it looked like Druid alpha female 569 had
chosen to den up the Lamar River at Cache Creek. The pack
now numbered three adults and eight pups from last year's
litters, who would now be considered yearlings. When I saw
569 later that spring, she had missing fur under her belly and
distended nipples, sure signs she had pups and was nursing.

The Slough females had similar indications they were
nursing. They were using a new den in a forested area along
Slough Creek where it was difficult to monitor them from
the ground or the air.

At the end of April, I got a good look at 302's left hind leg
and noticed his paw no longer had that twist to it. The broken bone must have mended in good alignment, although
that leg was still somewhat stiff when he walked.

In mid-June, the Druid alpha pair and some yearlings visited the pack's Chalcedony rendezvous site. 480 stopped,
looked back toward Cache Creek, and howled, probably to
contact 302 and other wolves who had stayed behind with
the new pups. We spotted at least four of those pups on the
slope of Mount Norris along with a yearling. Doug Smith

did a flight on June 28 and saw six Druid pups at the Cache Creek den.

Two days later, I spotted the three Druid adults, 480, 569, and 302, together with six of the eight yearlings a few miles west of the rendezvous site. Alpha female 569 was being especially playful. She wrestled with the yearlings and engaged in chasing games with them. I thought the mother wolf was in such good spirits because she was taking a break from caring for her pups. The other two yearlings were likely on babysitting duty with the pups, which by now would be about two months old.

By that time, the Slough wolves had moved their pups from their den site up Slough Creek to a new rendezvous site on a ridge between Slough and Lamar Valley. I hiked uphill to a knoll a mile from the site. From there I saw alpha female 380 and thirteen pups: ten blacks and three grays. The young gray alpha male was also there. A pond was nearby and the water source might have been a major reason for moving the pups to the new location. The count of thirteen pups suggested that there was more than one mother wolf as an average wolf litter is five or six pups.

On a later trip, I saw the thirteen pups playing with each other, mostly games of chasing, sparring, wrestling, and pouncing. 380 was bedded nearby and watching the play. One pup chased after flying insects. As the Slough alpha male was gray and most of the pups were black, I thought 302 was likely the father of some of them.

We saw the adult Slough wolves carrying off two black bear yearlings on the north side of Lamar Valley a mile or two from their new rendezvous site a few days later. There was a new bison carcass nearby and the wolves and bear family likely ran into each other at the site and fought. The wolves' success was unusual. Normally mother bears are highly effective in protecting their cubs.

I knew of five other cases of Yellowstone wolves getting cubs and all of them involved grizzlies. I witnessed one of these five cases myself, although I had only a partial view of the actual kill. The Mollie's wolves chased a mother grizzly with a new cub, which the mother got separated from. The wolves left her, easily caught up with the cub, and bit into it. Vegetation blocked my view but later a hiker found the cub's remains. The bear may have been a first-time mother and inexperienced in how to effectively protect her cub. Getting separated turned out to be a fatal error. I always assumed that wolves have a grudge against grizzlies because the bears often steal kills from them. A pack will do the dangerous work of killing a large prey animal such as an elk or bison, then a grizzly will come along and take over the site and prevent the wolves from feeding. Another issue is the frequent appearance of bears at wolf dens and the threat they pose to pups.

Matt Metz did a flight in late July and saw seventeen Druids at the Cache Creek den area, including seven pups. The three older adults were there, along with seven of the eight yearlings. The pack was much larger now, but still slightly smaller than the neighboring Slough pack, which numbered twenty-one. What matters more than the total count, however, is the adult membership, and the Druids had the advantage there: eleven to eight.

We were getting worried about the young Slough alpha male. He was often seen walking down the road in Lamar

Valley. His pack was spending more time in Lamar and they made several kills there, including one at the Druids' Chalcedony rendezvous site. So far they had not encountered the Druid wolves, but if they continued to return to that area a confrontation between the two packs was likely.

On August 1, nine Druid adults returned to that rendezvous site and found where the Slough alpha pair had done scent marking. 480 and 569 marked over that site and other places where the rival pack had been. 302 and a male yearling also marked those spots.

The Druids went back to the Chalcedony area on the tenth and did more scent marking. Some of the Slough wolves returned two days later, including their young alpha male. They sniffed around and seemed concerned. The Sloughs moved west toward their territory but frequently looked to the east, watching for the Druids. I got the sense that the young Slough alpha male wanted to avoid an encounter with the formidable 480.

In late August, the Slough pack moved their pups south of the road to Jasper Bench, a site they had first used in 2005. A female came in and did seven regurgitations to the pups, a record number. I got a count of eleven pups, two short of the high count of thirteen in mid-June. That area was about four miles west of the Chalcedony rendezvous site, where the Druids would likely soon move their pups.

I got up early on August 27 and left before dawn to head into the park. I had done that for 2,633 days in a row. Recently the big headline in sports was Cal Ripken Jr.'s record streak of starting in consecutive Major League Baseball games. It lasted for 2,632 games. I had now beaten him by one day. But I had things easier than he did. No one was throwing ninety-five-mile-an-hour fastballs at me.

In early September, the Slough wolves were at a new carcass in the Lamar River with several of their pups when four coyotes approached. A black pup took command of the situation and went toward them with its teeth bared. The little wolf intimidated the four coyotes into leaving without putting up a fight. Later three coyotes surrounded the pup and it charged at one, then the two others, driving all of them off. I felt that pup was likely going to be a future alpha.

I saw a new adult in the Slough group, a collared gray male. I did not get a signal from any local gray males in that direction, so I figured the battery on his collar had died. I did not recall seeing this gray with the Slough wolves before, but the alpha male seemed to be tolerating him.

Around that time, I noticed that a black Slough female known as Slant for her diagonal chest blaze was holding her front right paw off the ground as she traveled. The paw looked swollen and the toes were spread out far more than normal. Soon she was limping badly. After that, Slant held that leg up and resorted to traveling on three legs. Paw injuries are common among wolves. This one could have been caused by stepping on a sharp rock or being stomped by an elk. Most such wounds seem to heal, but we have had several Yellowstone wolves who limped for life after getting injured.

A major incident took place later that night. The young Slough alpha male was hit and killed by a car in Little America. His pack had injured a bull elk in the area, and he was probably looking for it when he ran across the road. He had a habit of walking down the road, and his death was a consequence of that casual attitude about traffic.

The new collared gray was still in the pack and he was now the only adult male. I expected him to take over the alpha position, but on the fourteenth, I saw him running from the pack, seemingly in fear. The other Slough wolves were in a big cluster, greeting each other and wagging their tails. Later I saw that alpha female 380 was bedded down next to a big collared black male. He must have come in and driven off the gray.

I pointed my tracking antennae at the black and got signals indicating that this was Agate yearling 590. He was likely a brother to the previous gray alpha male, the one killed on the road. Kira Cassidy figured out that the collared gray was a male from Idaho. He later joined the Beartooth pack east of the park—the group started by former Rose Creek alpha female wolf 9 in 2000. We were happy to see him because he would add some welcome genetic diversity to the local packs.

I remembered that we had seen 590 with the Sloughs last February, so the pack members already knew him. The pups took their cue from the adult females. When they saw them being friendly to the big new male, the pups apparently felt it was safe to be around him. Soon they were giving him play bows, invitations to interact with them. After that, all five adult females gathered around him. 380 marked a site by doing ground scratches with her hind legs, then 590 came over and did a raised-leg urination at the spot. The next morning, both did more joint scent marking and 590 played with the pups.

I FOUND THE Slough wolves at Slough Creek on September 18. They spotted two cow elk and ran toward them. One was slower than the other and the wolves targeted her. Several of the younger adults caught up and ran alongside the cow. A healthy elk can outrun a wolf, so this one likely had something wrong with her.

The cow ran into the creek, waded out to a deeper section of water, then turned to face the wolves swimming or wading toward her. 590, the new alpha male, came at her. She charged, reared up, and stomped on his back. Despite the blow, he circled behind her and bit into her hindquarters. 380 came at the cow and she stomped her front hooves on the wolf's back. The cow alternated between running from the wolves through the water and turning around, charging, and attacking.

The cow kicked down at 590 as he swam in and hit him on the back for the second time. The elk waded away from the wolves and they swam after her. Every time she stopped they would try to approach from the rear. But the cow would turn, charge, rear up on her hind legs, and try to strike down at a wolf with her front hooves. Then she would run off and repeat the sequence.

The turning point in the fight took place when a black grabbed one of the cow's hind legs and acted like a drag when she tried to break away. Other wolves rushed in and bit the cow on the sides and hindquarters. The cow managed to get into deeper water and the wolves had to swim to stay with her. But her wounds were slowing her down and they kept pace with her.

Alpha female 380 got out in front of the cow, turned, and got a good bite on her throat. Another wolf bit into the side of the elk's neck. Those bites and the previous ones weakened the cow to the point where she collapsed. Six wolves encircled her as she treaded water. There was no movement from the cow soon after that. Her body drifted to shallow water by a gravel bar. At least twenty Slough wolves gathered at or near the carcass.

TWO EVENINGS LATER, I counted seventeen Druids in the Chalcedony rendezvous site, including seven pups. The full count for the pack should have been eighteen. A black yearling was missing. The next morning the Druids were gone from Lamar Valley, but some of the Slough wolves were there, running toward an intact bison carcass. When new alpha male 590 arrived at the carcass a few minutes later, a feisty pup repeatedly snapped at him, defending its spot. The big male just stood there and did not retaliate. An adult female also snapped at him in annoyance. 590 did not react to her either and walked off without eating.

I was impressed by the new alpha's behavior. He was the biggest and strongest wolf at the site but did not respond in kind when the pup and small female got aggressive to him. He apparently respected that they had gotten to their feeding spots first and had the right to stay in place. Later he went back to the carcass and a pup lunged and bit him, but 590 still did not do anything. He was new to the family yet acted like I had seen 21 and other alpha males act, giving preference to pups and a female at a carcass.

I had been concentrating so intently on the wolves that it took me a while to realize this was not a bison carcass. The first clue was the lack of hooves. I studied the feet and saw

claws on them. Then I put my scope on the head and confirmed that this was a grizzly. The Slough wolves were eating
bear meat. People on the scene told me that the animal
was dead when the wolves arrived. Most likely this grizzly
had been killed by another, bigger bear. Later, as the pack
tugged on the carcass, the bear's head ended up propped
up in a lifelike pose. All the wolves were now hesitant about
approaching and acted like they were afraid that the grizzly
would bite them if they got closer.

Five days after that, I spotted eight adult Druids in Lamar
Valley. I heard that earlier they had been near a grizzly and
a carcass. That new carcass looked like it was another grizzly. No wolves or other animals were now near the carcass,
so John Kerr and I got our bear spray and walked out there,
carefully scanning for more bears. We reached the site and
confirmed that this was a second dead grizzly. It was an eerie
sight, for the rear half of the bear was buried and the front
portion, including the head, was sticking out of the dirt. That
burial indicated a grizzly had killed this bear and partially
covered it with dirt. Parts of the belly and rear end had been
eaten, most likely by the killer.

I called the park's bear biologist, Kerry Gunther, and he
came out with a crew. I guided them to the dead bear and they
found deep bite marks on the face of the grizzly. One tooth
had made a hole right between the eyes. That was an intimidating display of how much jaw strength a big grizzly has.

This was a nursing adult female, about ten years old and
around 275 pounds. Two of her canine teeth were broken off.
That might have happened when she tried to fight off the
bear that killed her. Most of her wounds were on her face.

We took that to mean she had not run from her assailant but stood up to what was likely a much bigger male grizzly who was threatening her cubs. Kerry found grizzly fur in the female's teeth, meaning she had gotten some bites into him.

Male grizzlies often try to kill first-year cubs so they can breed their mother and replace another male's cubs with their own. This mother fought to the death to protect her cubs, an impressive demonstration of courage. The big grizzly who killed her was likely the same one who killed the bear the Slough wolves had fed on. In the coming days I looked for a big bear with bite marks on his face but did not spot him. I definitely did not want to run into him.

On the last day of the month, all eleven Druid adults—the three older wolves and all eight yearlings—were at the Chalcedony rendezvous site. They howled and we heard answering calls from across the road that had to be from their pups. The two groups howled back and forth, and I could see that the adults looked stressed by the cars and people between them and their pups.

The adults moved east and soon were south of their den area north of the Footbridge lot. They howled, and the pups called back from the den forest. Mark Rickman spotted five pups up there. A group of people started to walk up toward those pups, making the situation much worse. I asked them to come back and they complied. Soon the pups crossed the road and reunited with the older pack members. Seeing them all together caused me to think that the Druids finally had enough members to effectively defend their territory against the Slough wolves and other rival packs.

16

The Oxbow, Agate, and Leopold Packs

I N 2007 A group of wolves formed the Oxbow pack and denned west of the Druids and the Sloughs, near Hell-roaring Creek. We found a high viewpoint where we could monitor the site, including the den entrance, which was a few yards from a small pond. The alpha female was 536 from the Leopold pack, a relative of 302 and 480. There were ten adults in the group.

On April 30, I saw pups come out of the den for the first time. The twelve pups alternated between crawling and walking in unsteady gaits. Several yearlings hovered over them. A few days later, I saw that the pups still had not mastered the art of walking and often stumbled or fell down. When they encountered a log, they crawled over it and fell off the back side. I guessed they were less than two weeks old.

I watched 536 as she went to the den and nursed pups. After she finished, she walked to the pond and got a long drink. Later the mother picked up a pup in her mouth and stuck it into the den. After the other pups came over and went in on their own, the mother slipped in. I later saw a second female nursing the pups. That meant that she must have given birth to pups of her own and they were likely in the den with 536's litter.

The next day, the pups approached the nearby pond followed by a yearling who acted like it was concerned that the pups were close to the water. It seemed to understand the danger of drowning. When a pup reached the edge of the pond, the yearling tried to pick it up but could not figure out the proper way to carry it. In frustration it pawed at the pup and knocked it over. The pup ended up running away from the pond.

By May 8, the Oxbow pups had good balance when walking around, even going to the edge of the pond and lapping up water. They were exploring the area and going farther from the den. They soon learned the trick of sneaking up behind an adult and tugging its tail, a prank I had seen many other pups play on older wolves in past springs.

One day the alpha male was bedded down near the pups. I heard a faint squeaking sound that seemed to be from a pup in distress. The male jumped up and ran to four nearby pups. I think a pup had nipped another one and the father wolf was running over to check on what had happened. That male was later collared, and he turned out to be a son of 21 and 42. We could see that he had a calcified lump on one of his legs. That indicated the bone had been broken but was now mended.

We eventually saw a third female nursing pups, meaning the twelve pups came from three separate litters. By mid-May, the pups were wading around in the pond. Laurie Lyman saw a pup bite the back of a smaller pup's neck and shake it like a piece of meat. A yearling saw what was happening, came over, and nudged the bigger pup with its nose, distracting it enough that it let the smaller pup go. The incident showed how a yearling will intervene if a pup is getting too rough with another one. The variation in pup size indicated that the three litters had been born at different times.

On May 15, wolf 536 moved the pups to a new den site. She got them to follow her to the far side of the pond, then picked one up by the belly and carried it west. 536 returned seventy-three minutes later and grabbed a second pup. I went to an overlook farther to the west to see where she was taking the pups. 536 soon came into sight, holding a pup by its rear end. I lost her going into a forest. I saw her carry a total of six pups west. I heard that when she returned to the original den to pick up a seventh pup, a yearling tried pushing two other pups west with its nose.

I spotted the alpha female with that seventh pup and saw a gray pup and two yearlings following her. The pup ran at a good pace after her, and she stopped and looked back frequently to check on its progress. The two yearlings ran along either side of the pup, looking like bodyguards.

Soon it got too dark to see, but people later told me that four pups were still at the original den. I went back the next morning and saw one of the Oxbow mother wolves walking around the area, checking for pups. After going in and out of the den a few times, she left and went west. Those last pups must have been carried to the new den during the night.

A few days later, Emily Almberg and I hiked to a hillside observation point where we could look for the new Oxbow den. We soon saw seven adult wolves bedded down at the likely site, over a mile away from our location. We saw 536 and the two other mothers, as well as the alpha male. Then we spotted eleven pups nearby. All the gray pups were there, but one black pup seemed to be missing. 536 nursed some of the pups, then all eleven pups went into the den. The distance from the original den to that new site was over two miles.

I ALSO WATCHED the Leopold pack that spring and summer, something I had done for eleven years. Their territory was west of the Druids and Sloughs and southwest of the Oxbow pack. The Leopolds were the first naturally forming pack to be established in Yellowstone after the reintroduction of wolf packs from Canada. All the earlier packs were either existing families or individuals put together in acclimation pens.

302 was born into the Leopold pack in 2000. In his younger years he had made the long trek—fifty miles round trip—between his pack's territory and Druid territory to visit pups he had sired before finally joining the Druids in 2004. That same year, the Leopolds had been the largest pack in the park with twenty-three members. This year they numbered thirteen adults with an as-yet-unknown number of pups.

As I watched the Leopold pack's den site one day late in April, I could see the face of alpha female 209 looking out from the entrance. She soon slipped inside to tend to the pups. Several bison walked into the area and one looked in the den entrance. It suddenly ran off, probably because the mother wolf growled at it. Other bison approached the site and also backed off. After those bison left, 209 came out of the den and shook dirt off her coat.

Other wolves went in and out of the den that day. Based on previous observations, we thought that four Leopold females were pregnant, and it looked as though all four of them were using that one den. Then two tiny pups that appeared to be about two weeks old came out of the entrance. The pups were just learning to walk and at times resorted to crawling around. Two yearlings carefully walked around the pups, seemingly afraid of stepping on one. One of the yearlings appeared to be fascinated by the pups and licked them repeatedly.

On a later trip to monitor the den in June, I saw the adult wolves repeatedly chase a black bear up a nearby tree, which is where it was when I first saw it. The bear came down three times and the wolves charged in and forced it back up each time. When it came down for the fourth time, on the far side of the tree from the wolves, it ran off. The pack chased it, but the bear got away. The Leopold adults had encountered the bear some distance from the den, but they probably wanted to make sure it left the area before it got any closer to the pups.

I HAD ALSO kept track of the Agate pack that spring and had seen that 472, their old alpha female, was pregnant. In late April, I saw the Agate wolves at a carcass when a grizzly came on the scene. The pack chased it, with former alpha male 113 in the lead. That was good to see, for it indicated the ten-year-old wolf felt well enough to harass the bear.

By early June, the Agates had moved their pups to a rendez-vous site at Antelope Creek, about six miles south of Tower Junction. The location offered good visibility of the pack to park visitors. We counted eight Agate pups, and later a ninth pup was seen, a sign that more than one female in the pack had given birth. 113, who did not travel much any-more, spent a lot of time resting here. He served the pack by watching over the pups when the other adults were out on hunts. The pups, who would have been his grandchildren, liked to follow the elderly wolf around in single file. They often harassed nearby adults for food or play sessions, but they did not bother 113 and may have understood that his age and injury limited what he could do.

As I watched 113, I thought he seemed to be accepting of his lot in life. He was the founding alpha male of the Agate pack and had successfully raised many generations of pups. His adult son 383, the new alpha, treated him respectfully, as did all the other adults. I saw 383 return to the rendezvous site with a full stomach. He went straight to his father with a wagging tail and both males licked each other in the face. I took that as a sign of mutual affection.

In early September, I was watching the two gray female yearlings in the Agate pack: 693 and the 06 Female. The huskier one, 693, went to her sister and lunged at her for no apparent reason other than to intimidate her. 06 might have been smaller than her sister, but she would prove to be smarter, and years later I would see her get revenge on 693.

That fall there was an unfortunate incident in the Agate rendezvous site. Three people walked out toward the wolves to photograph them, even though the area was closed to the public. They went right up to alpha female 472 and two pups. She bark-howled at them, a call that warns pups and other pack members of danger. Despite her obvious distress, the people moved closer. We radioed the law enforcement rangers and they dealt with the disturbance. Each member of the party had to appear in the court at park headquarters. Human disturbance caused the Agates to abandon that rendezvous site and many thousands of park visitors lost their chance to see the wolves there.

17

Conflict Among
the Packs

A S OF OCTOBER 1, the Druids and Sloughs were somewhat evenly matched. There were eleven adults and
seven pups in the Druid pack, and eight adults and
eleven pups in the Slough pack. If the packs clashed, the
higher count of adult Druids would give them the advantage.
Also, 480 was older and more experienced in battle than the
young Slough alpha and had a talent for combat strategy. The
Slough wolves regularly made incursions into Lamar Valley
and killed elk there. I figured 480 was waiting for the right
situation to move against the other pack and drive them out
of the valley.

The Sloughs had a more complicated situation than the
Druids for they had three other groups of wolves close to
their territory that matched them in numbers. The Oxbow
pack, with seventeen members, was making incursions

into the west end of the Slough territory. The Agates, with eighteen members, lived to the south. We had not seen the Unknown wolves for some time, but they were probably just north of the Slough territory.

This meant that the Druids had only one rival pack to deal with while the Sloughs were surrounded by four packs. On October 11, we got a report of a dead black pup on the west side of Slough Creek. I waded the creek with Erin Albers, a biologist with the Wolf Project, and we found the female pup with bite marks on her face and body. The Oxbow wolves had been seen to the west recently, so they were the prime suspects.

In mid-October, I spotted the Slough wolves on the north side of Lamar Valley. An outsider black male was nearby. I soon determined that it was a Druid male yearling. Since the breeding season was approaching, he was probably trying to find a mate in the Slough pack. But his family and these wolves were rivals competing for territory, and the Slough alpha male, 590, would likely consider him a threat.

I noticed that this young Druid male had a high level of social intelligence. He made friendly contact with the Slough pups and young adult females while avoiding 590, the only adult male in the pack. When 590 charged at the Druid, the young wolf immediately took off but ran just far enough to satisfy the big male. By late morning, I had seen the Druid male with six of the seven Slough females and many of the pups. 590 was mostly ignoring him by that time. When the two males later got near each other, the newcomer readily acknowledged 590 as his superior by licking his face.

The next morning, I spotted wolves at Slough Creek. Eight black adults were standing together on a hill. Since there were seven black adults in the family, 590 and six females, the eighth wolf had to be the Druid male. To successfully join a rival pack in less than a day was an amazing accomplishment. The young wolf had made friends with the pack's females and pups and seemed to reassure the alpha male that he was not a threat. Perhaps 590 was thinking that accepting a new male into the pack would help him cope with the threat of having four wolf packs surrounding him. Strategically, it was a smart move.

Later that day, the young black did a raised-leg urination and alpha female 380 marked over his site, indicating that she accepted him as a member of the family. Then 590 came over and the newcomer went to him in a submissive posture and licked his face. The big alpha male pinned the younger wolf and gave him holding bites that seemed to do no harm. Within the hour, both males were peacefully resting side by side like they were longtime friends.

A few days later, the Slough wolves, including their new member, were a few miles west of the Druids. The two packs howled back and forth but stayed apart from each other. By then it looked like a second yearling, a gray male, had left the Druids, probably to seek out a mate. That lowered the Druid pack count to sixteen.

I last saw former Agate alpha male 113 on October 15. By that time, his radio collar was no longer transmitting so I had to look for a collared gray that matched his appearance. By the end of the month, I reluctantly concluded that he had passed away. He would have been ten and a half years old, about eighty-two in human years.

Years later his collar was found miles from the Agate territory. As 21 had done at the end of his life, 113 must have walked away from his family, bedded down, and likely died of natural causes. I had known and admired him for many years. His greatest legacy would be his daughter, the 06 Female.

I SAW THE Druids chase elk one November morning in Lamar Valley. 480 singled out a large calf and followed it into the river. It soon stopped in deep water and turned to face the wolf. 480 ran toward it, leapt up, turned his head sideways, and grabbed the upper part of the calf's throat. Within thirty seconds, the calf collapsed into the water. If 480 maintained his tight hold a few seconds longer, the calf would soon be dead from lack of air. But when the wolf let go and ran off, I realized that the site was close to the road and people had stopped to watch the drama. Their presence was too much for 480 and he left.

In the end, it was 302 who saved the day. When 480 moved away from the crowd, 302 took over. He was much more used to the road and people than his nephew and had no problem with this situation. 302 ran in, went right into the river, rushed through the water to the calf, reared up, and grabbed its throat at the same spot where 480 had bitten it. The calf soon collapsed and died. 302 had failed his pack many times in the past, but on this hunt he was the MVP.

The local ravens quickly took advantage of the situation. The first one flew to the site within a minute of the calf's death and another one immediately joined it. Ravens are ever vigilant when wolves are in the area for if the pack makes a kill the birds will be well fed for days. In my research on ravens I found that at least one bird who lived in captivity survived to its sixty-ninth year. The time period between the killing of the last Yellowstone wolves and the arrival of the first fourteen wolves from Canada in the 1995 reintroduction was sixty-nine years. That means there could have been an old raven in the park who was alive back in 1926 and had regularly stolen meat when the original wolves made kills. On seeing the newly released wolves, it would remember the old days, follow a pack as it hunted, and get a free meal when the wolves got an elk. Other ravens would notice what that bird was doing and quickly understand how they could also get in on all the free food.

The day after that hunt, the Druids chased a group of bull elk and singled out one of them. The elk apparently was in poor health for the wolves easily caught up and ran alongside him. The two big males worked together: either 302 or 480 leaped up and bit into the shoulder while the other one grabbed the throat. The rest of the wolves bit the elk at different spots. The bite on the throat finished the elk off and the pack pulled him down.

These two kills showed that 302 was pulling his own weight during hunts and that his fear of bull elk was a thing of the past. 302 was over seven and a half years old and had already lived nearly twice as long as an average park wolf. I had been sure he was a lost cause and would never change his behavior patterns. But these two hunts showed that maybe he was capable of change, after all.

Right after that hunt, six wolves we didn't recognize came into the area and the Druids immediately chased them. They caught up with one, pulled it down, and bit at it. The Druids had caught a black pup and had it at their mercy. They could have killed it, but 480 stepped away, saw a gray adult from the other pack, and chased him, the other Druids following his lead. The black pup jumped up and ran off, seemingly uninjured. The bites must have been nips rather than full-force killing ones.

When we heard howling to the west of the Druids, I looked that way and spotted three adult wolves. One had a distinctive silver coat and another, likely the alpha male, was gray. The third wolf was black and looked like a young adult female. All three howled. The Druids howled back but did not charge at them. Later the black pup and two gray pups joined the three adults. We had seen that silver female and some of these wolves before in Lamar Valley but did not know who they were or where they were based. We called them the Silver pack after the coloring of the alpha female.

When things slowed down, I thought about the incident. I was impressed that 480 had that pup at his mercy but spared its life. I had seen former Druid alpha female 42 catch a pup from a rival pack years earlier, then let it go, and would later see other similar incidents, including a time in 2004 when the Druids caught a Slough pup and let it go. All that caused me to think that adult wolves seemed to have an instinct to stop attacking a pup from another pack once it went submissive. There had never been any conflict between the Druid and Silver wolves, so that may have been a factor as well.

The Druids went on another elk hunt two days later. They targeted a big bull who initially ran off, then stopped and stood his ground as sixteen wolves surrounded him. The bull kicked back at the wolves behind him and spun around to face one group of Druids, then another. He held his head up high, a sign of strength and defiance.

The wolves seemed respectful of that vigorous defense. They stayed just out of the reach of his hooves and antlers. When the bull trotted off, they followed, but did nothing when he stopped and faced them. Apparently deciding that this elk was too strong for them, alpha female 569 called the hunt off. She walked away and the other wolves followed her lead. The Druids went west, found another bull elk, and surrounded him as he confidently stood his ground. Also judging him too vigorous to take on, the pack walked off.

Earlier we had seen the Slough wolves feeding on a bull elk carcass to the west, and the Druids were now heading that way. Eighteen Slough wolves were still at the carcass, but many were pups. The Druids numbered fifteen and also had a lot of pups in their group. 590, the new Slough alpha male, must have seen the Druids approaching for he left the carcass and went west with most of the other pack members following. A single young adult female lingered, apparently unaware of the danger.

I scanned east and saw the Druids charging that way with 480 in the lead. The Slough female ran off. The Druids then saw a nearby black Slough pup and charged at it. The pup ran to the river and disappeared in a low area with the Druids just a few lengths behind it. They ran into the depression and it looked like they were biting something. I could not see if the pup was going submissive or fighting back. After just a minute the Druids walked off. They caught sight of two other Slough wolves and chased them, but the Sloughs got away.

The Druids went back to the attack site by the river and seemed to be nipping at something on the ground. They soon left, but I still could not tell how that pup was doing. The wolves spotted the nearby elk carcass the Sloughs had been feeding on and ran to it. Slough pups, which were still in the area, howled. Distracted by the carcass, the Druids ignored them.

Slough alpha female 380 arrived, traveling alone. She was at great risk of being seen and attacked by the Druids, but she continued toward the howling pups. I saw four Slough pups running to her. As soon as they reached her, she trotted off to the west, leading them away from the rival wolves. The Slough pups howled, probably something 380 did not want them to do because it gave away their position. The Druids howled back. Alerted by the howls, another Slough pup found 380 and the group moved off. By that time, the Druids had bedded down and the conflict was over.

We scanned the area and saw seven live black Slough pups. A recent count of nine black pups in the pack was down from eleven. Accounting for the one attacked by the Druids, that meant that one more black pup was in the vicinity. It was later seen north of the road. We watched the low area by the river where the Slough pup had been caught by the Druids and could not see anything. Had it died or gotten away, or was it hiding? No ravens flew into the area, a sign that it was not dead.

It snowed the next morning and the temperature dropped below freezing. If the Slough pup was injured, the snow and cold would hinder its recovery. The Druid and Slough signals were all miles away from where we had last seen that pup. Later that morning, two ravens landed and pecked at something out of sight in a low area. Their presence indicated that the pup had died of its injuries, probably due to blood loss. The pup's fate contrasted with how the Silver pup pinned by the Druids had survived. Perhaps the Slough pup had fought back more than the Silver pup, and that was what caused the Druids to bite it more times, possibly in a vital area. There was also the issue of the history of aggression by the Sloughs against the Druids, while the Silver pack had never bothered them.

Wolf Project volunteer Jerod Merkle and I walked out to the dead pup. There were bite marks on its back, neck, and hindquarters. I thought about how the Slough wolves had killed Druid female yearling 375 in February of 2005. Whether the Druids intended to kill this pup or not, his death evened out the score between the two rival packs.

In mid-November, I saw all sixteen Druids traveling west through Lamar Valley. They picked up a scent trail, probably of the Slough wolves, and got excited. 302 led west at a fast pace. When the Sloughs howled from north of the road, the Druids looked that way and 302 led north and crossed the road. The alpha pair and eleven younger wolves followed, but some of the pups stayed on the south side of the road.

I rushed uphill to an observation point and got in place just in time to see the Druids running downhill. They looked like they were chasing something. The younger wolves were leading and a big male, either 302 or 480, was last in line. The sage was so thick I could not see what they were chasing.

The lead wolves abruptly stopped and attacked something in the brush and the other pack members ran in and gathered at the spot. I got a glimpse of a black wolf on the ground. At that point, some of the wolves walked away. The last one left a few minutes later. Within a minute or so, ravens landed and pecked at what by now must be a dead wolf. The Druids went north, and I lost sight of them.

Later a Wolf Project crew hiked up there and found the wolf. It was the young Slough adult female known as Slant, the one with an injured paw. That must have slowed her down and allowed the Druids to catch her. She likely fought back and refused to submit, causing the Druids to continue biting her. That more than evened the score between the two rival packs regarding fatal attacks, but the Druids had not yet driven the Slough wolves out of their territory, which meant that the Druids were still not safe.

I noticed that 302 was an active participant in the Druids' campaign to force the Sloughs out of their Lamar Valley territory. He seemed to be a different wolf from the one that had snacked and napped as the battle of Mount Norris raged a safe distance away from him. 302 was the most unorthodox wolf I had ever known. He had often been a liability to his pack, but now, against all expectations, he was becoming a valuable contributor. I was not sure what had triggered the change, but I was glad to see 302 acting more like a traditional adult male wolf who actively supported his pack in two critical areas: hunts and territorial disputes.

AROUND THE TIME the Druids were clashing with the Sloughs, an uncollared young male wolf with a light gray coat showed up in Lamar Valley, near the Druid den forest. He was destined to be an important player in the coming months.

The newcomer did a lot of howling, probably because he was looking for a mate. The Druids spotted the new wolf and ran at him, but he raced off and eluded them. Later a gray yearling Druid female met up with the gray male. She greeted him in a friendly manner by touching her nose to his, wagging her tail, and doing play bows. Their interaction was interrupted when the rest of the Druids charged at him and the gray male had to flee once again.

Another stranger showed up soon after that, a gray male with a dark coat. He met up with a black Druid female and went through the same greeting behavior as the gray with the light coat had with the gray female. Alpha female 569, who was probably the black's mother, soon appeared and stalked the newcomer. At first he tucked his tail and moved off, then he went to her and the two wolves had a friendly meeting. When 480 came on the scene and charged at the outsider, the dark gray ran off, then stopped. 480, followed by two pups, went up and sniffed the newcomer. The dark gray stood his ground until other Druids approached, then decided it was time to run away again.

By that time, the young Druid black male who had joined the rival Slough wolves back in October had worked his way up to the second-ranking male position in the pack. He would soon be collared and given the number 629.

AS WE GOT into December, I realized that the six yearlings remaining in the Druid pack were all females who would be able to breed in the coming months. The two outsider gray males were still in Lamar Valley, so they were well positioned if they wanted to mate with those females or run off with some of them to start new packs. In the Slough pack, alpha male 590 was showing interest in three of the pack's females. Other females favored 629.

The two gray males seemed to tolerate each other and continued to stay near the Druids. 480 chased Dark Gray into a low area one day. It looked as though he was beating up the younger male, but when wolf watchers saw the young wolf later, he did not seem significantly injured. The incident caused me to think that 480's intent was to warn him to stay away rather than kill him.

Light Gray had a patch of missing fur under his belly that looked like it was caused by mange. That was troubling for if he was infected, he could easily spread the mites to the Druids. I watched as Light Gray licked the face of the Druid alpha female, who tolerated his attentions. 480 approached and the young male gave him a submissive greeting. 480 did not act aggressively to him, indicating that he also had some tolerance for the young outsider. Later 302 came on the scene and chased Light Gray for a half mile, but the young wolf was too quick for him.

I saw the Druids attacking a sickly bull elk that was missing a patch of fur on his back, a sign he had scabies, an infestation of mites similar to mange in wolves. He ran to a creek and stood for hours in the cold water. The Druids stayed nearby and monitored him. After a while, the bull collapsed and just his head and back were out of the water. In such a weakened condition, he could no longer keep his nostrils above the water and drowned. The wolves fed on his remains. His death might have saved other elk from getting infected.

Scabies in elk is caused by mites in the genus *Psoroptes*, while the mite *Sarcoptes scabiei* causes mange in wolves. I checked with two experts, and their understanding was that mites that specialize in wolves do not normally infect elk, and the elk mites are not known to cause mange in wolves. Cases of scabies in the general Yellowstone region have been documented since at least the 1950s, although I have rarely seen elk with scabies in the park.

ONE DAY IN the second half of December, Druid alpha female 569 was in a particularly playful mood. She frolicked with the younger females, running circles around them and playing chasing games. Later she bounded through the snow to 480 and jumped around him, then rolled on the ground. When he bedded down, she lay down on top of him. 480 got up and jumped up and down in front of her and they rolled on the ground together. After that they sparred with their jaws, then the big male gently put a front paw on her shoulder. I wondered if they were in good spirits because it was now winter, a time when cold weather and deep snow weaken elk and make it easier for wolves to hunt them successfully.

The following morning, 302 led the Druids to a bedded bison bull, but something seemed wrong with the animal. Then I noticed a lot of ravens in nearby trees and one on the bull's back. When I took a close look, I realized he was dead. The pack fed there for many days. I had always figured that 302 was shrewd and felt he had likely spotted the ravens and wanted to see what was causing them to congregate there.

571, who had helped 480 drive off the Mollie's wolves when she was a pup, was the only one of the six Druid female yearlings that had a collar. We had to devise ways of identifying them in our field notes, and we usually chose a feature that related to their appearance. Two of the uncollared females were gray and one had striping high on her sides. We called her High Sides. The other had striping lower on her sides, so she was Low Sides. There were also three uncollared blacks. The one with a white vertical chest blaze became known as White Line. Two other blacks had horizontal bars on their chest. One bar was bright and the other dull, so they were Bright Bar and Dull Bar.

The two gray males continued to stay in Lamar Valley and usually were not far from the Druids. As we got close to the mating season, all six young females sought out those males and interacted with them. Since those sisters were too closely related to 302 and 480 for mating, the arrival of the two new males meant the sisters had potential mates who could help them start new packs.

IN LATE DECEMBER, the Druids chased a bull elk and 302 became the star player once again when he leapt up and grabbed the bull's throat. Yearling 571 ran in and bit into the side of the bull's neck. She and 302 worked together to wrestle the elk to the ground while other wolves attacked the hindquarters. Soon the bull was dead. In this hunt, 302 showed not only that he had overcome his phobia about bull elk but also that he was an effective leader and role model for the younger pack members.

302 had become the most famous wolf in the world by that time because of a recent Bob Landis television documentary featuring him. *In the Valley of the Wolves* was shown in America on the PBS series *Nature* and in the rest of the

world on the National Geographic channel. I got a telephone call from the office of Montana senator Max Baucus and was told he had watched the documentary and wanted to come to Yellowstone to see 302, so I arranged to meet him in Lamar Valley. When he arrived, he got in my car and we drove a few miles east and found the Druid pack. I set up my scope and showed him the wolves. The senator was especially captivated when I adjusted the scope and put it on 302. Despite getting older, the wolf was still strikingly handsome and unquestionably charismatic.

We soon saw 302 chasing off Light Gray. I spent six hours with Max, who kept asking for more stories about 302 and the other wolves. As new people drove up and got out of their cars to watch, he invited them over and told them about 302 and the Druid wolves. The senator had a naturally friendly and genuine manner and he was clearly enjoying talking to regular people about nonpolitical matters.

Bob's documentary was such a hit that he was asked to do a follow-up that told more about 302's life story. *The Rise of Black Wolf* was shown repeatedly all over the world on the National Geographic channel in 2010. We heard that it was the highest-rated wolf documentary they had ever presented. He did not know it, but 302 was an international television star.

I got a call from a producer of reality television shows around that time. He wanted to do a series about me in Yellowstone. I had no interest in being on television and knew that those shows often fake story lines for dramatic purposes, so I turned him down. After the call I thought I should have suggested that they sign up 302 to star in the next season of *The Bachelor*. I could see him in that series, along with a dozen or so young female wolves competing for his affection.

One day in late December, Light Gray was with two of the young females and they were being very affectionate and playful with him. A third young female came over and treated him the same way, licking his face and pawing at his shoulder. The three sisters squabbled a bit among themselves, each one wanting the male to pay attention to her. I looked around and saw that 302 and 480 were staring at Light Gray. He noticed their glare and moved off in a crouch, his tail tucked. Then the two big males charged, with 302 in lead. Satisfied that the outsider male was fleeing, both Druid males stopped.

Soon after that, 302 went after Light Gray by himself. The young male got bogged down in deep snow. The much bigger and stronger 302 plowed through the snow, bit into Light Gray's rear end, and threw him down. Then 302 slipped in the snow and fell. Both males jumped up and fought, and I could see them biting each other. 302 pinned the younger wolf and bit him. The young wolf squirmed out from under 302 and ran off, but 302 caught him and threw him down into the snow again.

When they fought this time, the young male bit 302 in the face. The next moment, other Druids ran in and Light Gray had to let 302 go and flee. Laurie Lyman got a close look at Light Gray and saw three bloody wounds on him. I spotted him bedded down licking a wound on one of his front legs soon after that. There were also bites on one of his hind legs and on his right side.

I then looked at 302. Fur was sticking out every which way on his coat. He seemed stiff and walked slowly, like he was in pain. I saw him wipe the side of his face with a paw, where the other male had bitten him. One of the young Druid females came over and licked his face, then a pup came over and sniffed 302's rear end, which suggested he had been bitten there as well. Later he licked a front leg and a hind leg. He had been bedded down, and when he got up, I saw blood on the snow.

That was the first time I had seen 302 refuse to give up while fighting another male, even when his opponent had given him what must have been a very painful bite to the face. The wolf continued to surprise us with how he was changing. 302 had known three greatly accomplished alpha males in his long life: his father, wolf 2; his uncle, 21; and now his nephew, 480. They were all aspirational role models on how proper alphas should behave and it looked like 302 was finally beginning to emulate them.

Can a tiger change his stripes? Can a person or a wolf change his basic character in his later years? Against all my expectations, it looked like 302 was starting to do just that.

Longtime alpha pair of the Druid Peak pack: male 21 (left) and female 42 (right) with beta male 253. After the death of the alphas in 2004, wolf 253 inherited the alpha male position. He was a worthy successor to his famous father, 21. **Doug Dance**

Young male 302 showed up in Druid territory in early 2003, got several of 21's daughters pregnant, then seemed to abandon them. But 302 repeatedly returned to the Druid territory that summer to visit those females and his pups. 21 and 253 drove 302 off whenever they saw him. **Doug Dance**

Wolf 302 (left) returned in 2004 with his nephew 480 (right) to challenge 253, who defeated them and retained his alpha male position. When 253 later left the Druids, 302 took over as the new alpha. But he soon failed in his responsibilities and 480 had to assume the alpha position. Female 569 is in the middle. **Ray Laible**

The Slough Creek pack often invaded Lamar Valley, the home of the Druid wolves. Since the Sloughs were a larger pack, the Druids had to retreat to the far reaches of their territory. Then a third pack came on the scene and changed the balance of power. **NPS/Dan Stahler**

Druid alpha female 286 with five young black pups in 2005. This aerial shot was taken from over 500 feet above the wolves with a high-power telephoto lens. Some of the pups are looking up at what to them probably seems to be a big bird. **NPS/Dan Stahler**

A group of seven Druid pups, likely sired by both 302 and 480, playing together. The most common games played by pups are chasing and wrestling. Both activities train pups for hunting prey animals and fights with rival packs. **Doug Dance**

Older pups scattering when a big bison charges at them. Due to their size, bison are very dangerous animals for wolves to hunt, but bison die of natural causes on a regular basis and wolves make good use of their remains.
Ray Laible

The pups have joined 302 (above the two bull elk) and 480 (confronting the bull to the left) on a hunt. Some pups are charging in while others are hesitating. The bull elk near 480 has uneven antlers, a possible sign of poor health. 480 picked that bull to confront, and the pack pulled him down.
Bob Weselmann

When wolves do all the hard, dangerous work of a successful elk or bison hunt, they usually have to put up with coyotes who try to steal some of the meat. **Bob Weselmann**

Ravens also steal meat at wolf kills, but wolves sometimes find animals that died of natural causes by spotting ravens circling the site. Ravens hang out at wolf dens and pups get used to having them around. **Bob Weselmann**

In 2008 there were six young Druid females, and they became enamored of two male suitors. In this shot, two females are running alongside the two males: Light Gray (far left) and Dark Gray (second from the right). 480 and 302 took an instant dislike to the intruders and tried to chase them off. **Doug Dance**

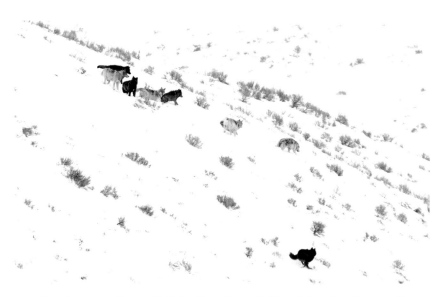

302 charging in to chase off Light Gray from six Druid females. Years earlier 302 was the renegade intruder who tried to mate with young Druid females and was chased off by 21 and 253. Now he was the one driving outsider males away from the family's females. **Doug Dance**

The 06 Female. Her name came from the year of her birth in the Agate Creek pack. 06 was the most independent and self-sufficient female wolf I have ever known. Many young males tried to court her, but like a Disney princess, she rejected them all, preferring to live as a lone wolf. **Gary Zylkuski**

In late 2008, wolf 302 took off to the west in search of females. At eight and a half, he had lived nearly twice as long as an average Yellowstone wolf. If he was ever going to start his own pack and be a proper alpha male, it would have to be now. **Doug Dance**

Several Druid yearling males joined 302 (center front) and they met up with a group of Agate females including 06 (right of 302) and her sister 693 (far left). The young males were drawn to 06 and ignored her sister. 302 acted as a chaperone and protected 06 from the clumsy advances of the yearlings. **Bob Weselmann**

302 started his own pack with those young Druids and three Agate females. He had failed early in life when he was briefly the Druid alpha male. He had a second chance to prove himself, but given his advanced age, the odds were very much against him. **Jimmy Jones**

PART 5

—

2008

18

———

The Two
Interlopers

A T THE START of 2008, there were sixteen wolves in
the Druid pack: the alpha pair 480 and 569, 302, six
female yearlings, and seven pups. Low Sides and High
Sides seemed to be the highest ranking of the six yearling
sisters. Light Gray and Dark Gray were still nearby, trying to
draw off some of the young females.

Probably because the breeding season was approaching,
480 frequently dominated 302 by pinning him. In 2005 wolf
302 had bred Druid alpha female 286 before 480 could stop
him. The following year, 302 had mated with the current
alpha female, 569. Normally an alpha male prevents lower-
ranking males in the pack from breeding with females in the
group, but 302 had a talent for getting what he wanted in
the mating season. 480 would have to vigilantly monitor 302.
One day, when 569 got near 302, alpha 480 jumped between
them and made sure nothing happened.

In January I saw Light Gray with four Druid pups. He had been around the pack for so long that the pups were used to him. Light Gray played chase with one pup. He and his group of pups then crossed the park road and one of the pups found an orange traffic cone and carried it around. Soon it stopped and chewed on the flexible plastic material. Then another pup came over and sparred with the first one over the cone. Watching those wolf pups made me think how dogs, especially young ones, love to play with new objects and chew on them, a trait that comes from their wolf ancestors.

I studied Light Gray and saw he had a lot of scars and wounds, some of which he probably got during his fight with 302. The wolf had been in Druid territory for about five weeks. At first, I assumed his plan was to run off with one or more females and start a new pack, but I now thought his goal was to join the Druid pack, the same thing 302 and 480 had done years earlier. I had to admire his persistence.

He still was in a rivalry with Dark Gray over those females. I saw Low Sides have a friendly meeting with Dark Gray and both flirted with each other. Right after that, Dull Bar did the same thing with Light Gray. Later she went to Dark Gray and consorted with him.

Dark Gray tried to mate with Dull Bar on January 16, but she was not in season and turned his advances into play sessions. Light Gray charged in with his tail raised and the other male ran off. Dull Bar went to Light Gray, and they played and interacted for some time. If that dominance by Light Gray over Dark Gray continued, it would give him an edge with the Druid females.

January 17 was the fifth anniversary of my first sighting of 302, back in 2003. That day one of 21's daughters went to

302, but her father ran over, chased the newcomer, caught him, and beat 302 up. 302 repeatedly came back that mating season. 21 was seven and a half years old when he was chasing off 302. 302 was now also seven and a half and doing exactly what 21 had done at the same age, driving off a young male from the pack's females. He had come full circle in his life.

Later in January, we had the coldest temperature in Lamar Valley so far that winter: minus 47 degrees Fahrenheit (−44 Celsius). I watched the sixteen Druids that morning and they seemed immune to the cold thanks to their thick winter coats. I was not as tough as the wolves and had to go back to my car to warm up.

480 tried to mate with 569 the next day, but she was not receptive to him. Later she went to the bedded 302 and pawed at him, a sign she still preferred him over the alpha male. 480 came right over and got between them.

A gray Agate pup was hit by a car just west of Tower during the night in late January. The wolf survived but her tail was injured and one of her hind legs was broken. I saw her a few weeks later with her family. She held that leg off the ground and traveled on her three good legs. The lower half of her tail eventually fell off.

At the end of the month, 302 got up on 569 and tried to get in a mating tie with her. 480 ran over and knocked him off. After 480 walked away, 302 got up on her once more. Looking back, 480 saw what was happening and shoved him off again. 302 then acted very subordinate to his much younger nephew. He would approach 480 in a crouch and roll on the ground under him, which was probably a calculated act to appease 480, for he went right back to 569 for another mating attempt.

The two older males did not seem interested in mating with the younger females, probably because they were closely related to them. Those six females had all been born in 2006. I had seen 480 breed alpha female 529 and 302 mate with 569 that year, so each would have been the father of at least some of these females.

All the Druid females came into estrus that spring. Since 480 was guarding 569 so closely and the younger females were of no interest to him, 302 had little chance of doing any mating in the Druid pack. He would have to leave to seek out other females.

480 mated with 569 three times on February 3. The next morning, I failed to get 302's signal in Lamar Valley. I went west and picked up his frequency at Slough Creek. He had gone there during the previous breeding season. That was a dangerous endeavor for there was a lot of bad blood between the two packs, and the previous year he had returned with a broken leg. Apparently he did not find any Slough females, for 302 was back in Lamar Valley on February 5.

302 constantly surprised us by behaving in unexpected ways. I later saw him with Dark Gray, and the two seemed to have come to an understanding. The younger male was in a submissive crouch as 302 wagged his tail at him. The gray rolled on the ground and pawed up at the old male, then got up and licked 302's face, acting like a pup. 302 lunged and pinned him a few times but did not bite. Each time he ended the interaction by wagging his tail.

I tried to figure out what might be going on in 302's mind. Perhaps by that point he knew that he was not going to mate with the younger Druid females because he was so closely related to them and he had grown used to the presence of the two outsider gray males. The submissive behavior of Dark Gray showed that he viewed 302 as a higher-ranking male and would pose no threat to him.

One day two of the young Druid females joined both gray males, which was unusual because the males were rivals for the females' attention and usually kept apart. 480 and other Druids soon charged in to break up the group. On some of the charges, 302 took the opportunity to stay behind and go to 569. That meant that 480 had three males to monitor. He would drive off the younger males, then run back to check on what 302 was doing. He could not trust him to be alone with the alpha female.

Back in the Agate pack, Erin Albers saw alpha female 472 mate with a black male who had come into her territory from another pack. Since the new alpha male, 383, was 472's nephew, it was better for her to get pregnant by an unrelated male. They had bred the previous year when 383 was getting established as the pack's new alpha male, but that might have been because of a lack of options.

302 LEFT THE Druids on February 10 and his signal once again came from the Slough territory. He was back with his pack three days later. I suspected that he bred one or more Slough females during his trip. On Valentine's Day, Light Gray mated with Bright Bar.

Doug Smith darted and collared a gray female Druid pup at the west end of Lamar in mid-February. She weighed eighty-one pounds, big for a pup that age. After checking the pup for health issues, the crew flew off to let her recover without any humans nearby, and I monitored her from the road to make sure she was all right. As I was keeping an eye on her, I spotted the Silver pack alpha pair a few miles away and was concerned they might find the pup while she was unable to defend herself or run away.

About two hours after the pup was darted and tranquilized, she had her head up. Soon after that she was up and walking around. She probably heard howling from the Silver wolves because she moved toward them, apparently thinking the howls were coming from her family. Then suddenly she turned around and ran off—she must have realized the wolves were not Druids.

The Silver alphas spotted her and charged. When the big Silver male caught up with her, he pulled her down and nipped her a few times. His mate arrived and, probably deciding she was no threat, did not touch the pup. At that point, the young Druid jumped up, stood her ground, and threatened the big male wolf with open jaws, an impressive display by such a young wolf. The Silver female turned around and walked off and the male followed. The pup did not appear to have any wounds from the male's nips, so her thick winter coat must have protected her. She rested for a while, then trotted off.

That incident was especially intriguing since I had seen 480 catch a Silver pup the previous fall and spare its life. In this case, the Silver alphas spared the life of a Druid pup. Soon after the Silver wolves left, I saw the Druids to the east. 302 started a howl and the others joined in. The pup must have heard them for she went that way. By that time it was getting dark, so I headed in. Early the next morning I saw that the newly collared pup was back with her family.

BREEDING SEASON WAS reaching a peak, and 302 went on another walkabout on February 16. We saw him in Agate territory at the west end of Specimen Ridge near a black wolf that must have been an Agate female. Soon we spotted thirteen other Agates in the area. I heard of an earlier sighting in the same area when Agate alpha male 383 chased off a black wolf that was probably 302. The two incidents showed that 302 was interested in getting to know the Agate females.

Back in the Druid pack, the perseverance of Light Gray and Dark Gray paid off. Light Gray twice tried to mate with Druid female 571. Soon after that, he got into a tie with one of the uncollared gray females. The next day, Dark Gray also mated with an uncollared gray. 480 ran at the pair when they were still tied together. Dark Gray frantically tried to break the tie but was still connected to the female when 480 arrived. He pounced on the young male, ending the mating. 480 had Dark Gray on the ground and was biting him, but the young male jumped up and fought the Druid alpha. The older, more experienced male was too much for him and Dark Gray ran off.

During the mating season, Light Gray bred at least two of the Druid females and Dark Gray at least three. One female tied with both males. I later told a group of people in Lamar Valley about how the six Druid sisters had to share two boyfriends, and an eight-year-old girl said, "That's never good."

Children always loved hearing stories about 302. I felt that was because he had such a big personality. I did a talk for a group of fourth and fifth graders from Harlem who were taking part in the park's Expedition Yellowstone educational program. They eagerly listened to my 302 stories and especially liked the ones about the times he messed up. I later gave a talk to another school group, and a young girl named Andrea gave me a drawing of 21 and 302. She encircled 302 with little hearts. That picture is hanging on my wall as I write this.

By the end of the 2008 mating season, 302 had gone into the Slough, Agate, and Leopold territories in search of females. Dan Stahler flew over 302 one day and saw that he was with five uncollared gray wolves, probably all young females. I later saw him with a gray Agate yearling that looked like the 06 Female. She averted her tail to him. He sniffed her but must have figured out that she was not quite ready, so he stepped away and bedded down. 302 was back with the Druids on the twenty-third. Light Gray and Dark Gray continued to stay in Lamar Valley after the breeding season.

ON THE FIRST day of March, I saw 302 and a black yearling following a cow elk. The elk must have been in poor condition for she was getting bogged down in deep snow and could not manage anything more than a fast walk. Wolf paws spread out when the animal puts weight on them, creating a snowshoe effect. 302 had especially big paws so he could often walk or run on the surface of a snowfield where much heavier prey animals sank into snow at every step.

302 walked up behind the cow and grabbed a hind leg. The yearling got a holding bite on her right shoulder. As the yearling maintained her grip, 302 let go, moved forward, and bit into the front of the elk's throat. Four more Druids ran in, including three pups, and the six worked together to pull the elk down and finish her off. The cow was in such a weakened

condition that she barely struggled. She was dead within four minutes.

The kill took place close to the road. As cars stopped and people got out, most of the wolves backed off. 302 and one pup stayed and fed, then that pup left. Seven other Druids were uphill, waiting for the people and cars to leave. Erin was working with me that day and she put up No Stopping signs. That cleared out the cars, but then I saw that three people had parked down the road and were walking toward the carcass. That was too much even for 302 and he left.

We called in the law enforcement rangers to help us deal with the situation, and they dragged the carcass farther from the road so the wolves would feel comfortable going to it. Erin examined the cow as the crew was getting ready to move her and saw that one of her legs had plunged through the snow so deeply that the surface of the snow touched the cow's belly. Her teeth were worn down to the gumline, indicating she was very old. With such poor teeth she would not have been able to feed properly. All those issues made her an easy kill for the wolves.

A few days later, the Slough wolves were on a fresh kill when Light Gray appeared. A yearling male and alpha male 590 charged at him and pulled him down. Light Gray fought back, broke free, and ran off. The two males and six other wolves chased him into some trees. I was thinking that Light Gray was taking too many chances with the local alpha males. He survived this attack, but if he continued to intrude into their territories he might eventually be killed.

March 21 was the thirteenth anniversary of the release of the original reintroduced wolves. I had been out in the field for 81 percent of the days the wolves had been roaming free in Yellowstone. We were in the middle of a Golden Age for wolf research. During those years we had so many wolf sightings and were seeing so much intense behavior, some of which had never before been observed, that I often had no time to eat my lunch, even though it was within easy reach.

I soon saw the Agate pup that had been run over by a car. She was back with her pack and doing well. She could put weight on her broken hind leg when she walked but held it up when she ran. I watched her having a tug-of-war with another pup over an elk leg and saw that she was strong enough to gain possession of it. Later her leg healed to the point where she had only a slight limp. She was back to normal thanks to the bone-making process Mark Rickman had described to me. The loss of half her tail did not seem to bother her. She could still wag it.

I saw the Slough wolves chase and fight a big bull elk in late March. There were three pregnant female wolves in the group, and all of them stood off to the side while other pack members battled with the bull. Those females probably instinctively knew to avoid dangerous situations during the later weeks of their pregnancy. The other wolves finished off the bull. Wolf Project biologists later examined him and found that he was an old animal in very poor condition.

LATE IN THE month, we got signals from the Druids at their traditional den in the forest north of the Footbridge lot. They had not used the site since 2004 after frequent incursions by the Sloughs. Since the Druids were now the dominant pack, we hoped they would reuse the site. If they did go back to their original den, it would be the ninth year they had used it.

All six of the young Druid females looked pregnant. Since none of them bred with 302 or 480, the sires of their pups would be Dark Gray or Light Gray. 569 was also pregnant. With sixteen wolves in the pack, that meant there would be nine members supporting the seven females: 480, 302, and the seven pups, who would be yearlings by the time the next generation of Druids was born.

One morning the Druids spotted a herd of bighorn sheep in a high-elevation area near their den. I had never seen Yellowstone wolves get a sheep because bighorns almost always stay in or near steep cliff areas. Five bighorn ewes saw the wolves and ran to a rock outcrop. Two of the pups charged at them and a yearling followed. The sheep panicked and ran back and forth on the cliff.

Then the sheep made the mistake of leaving the safety of that cliff and making a dash for another cliff a short distance away. Two of the sheep got bogged down in deep snow. Their narrow hooves are perfectly designed to climb narrow ledges on rock walls but give them little support on snowfields. A female yearling reached one of the sheep and pulled her down, and the rest of the pack ran in and quickly finished her off. I recall that none of the older adults participated in the sheep hunt, probably because they had learned from experience that it was almost impossible to get a bighorn. An advantage of having a lot of younger adults in a pack is that they will try things the older ones might not bother with. They usually fail, but sometimes they get lucky and score a big meal for the family.

As the 2008 wolf mating season ended, I thought a lot about 302. He was now nearly eight years old and had lived almost twice as long as the average Yellowstone wolf. Next year he would be the same age as 21 when he died. If he ever was going to take on the risk and responsibility of starting his own pack, 302 would have to do it soon or he would run out of time. I never told anyone, but I was secretly hoping he would end up as the alpha male of his own pack. What a story that would be.

19

The Tragedy of
Light Gray

B OB LANDIS SAW the Slough wolves digging out the
Natal Den at Slough Creek in early April. The pack had
not used that den since 2006, the year the Unknown
wolves had laid siege to it and all the pups were lost. The den
sites they had used the last two years were out of sight to us,
but this year I would be able watch the den every day. I saw
big alpha male 590 go to the opening, look in, and wag his
tail like he was greeting a female inside. As far as we could
tell, three of the females were pregnant: 380, 526, and one
we called Hook because she had a twist in her tail.

Although it was officially spring in Yellowstone, it was
often still cold in the predawn hours. On April 2, the low in
Lamar Valley was minus 14 degrees Fahrenheit (–26 Celsius).
I got 480's signal at the Druid den forest that morning, but
not 569's, which probably meant she was underground. She

was in the den the next day as well, so it seemed that now that the pack was larger and in control of that part of Lamar Valley, 569 had decided to return to the family's traditional den. The pups from last year were now yearlings and could take on new responsibilities such as babysitting and helping to provide food for a new generation of family members.

Doug McLaughlin, who managed the Silver Gate Cabins Resort in our little town, was one of the best wolf spotters we ever had. That skill is hard to define but appears to depend on a person's ability to notice details on the landscape. Hunters and bird-watchers, for instance, nearly always make good wolf spotters. Doug had owned a large thoroughbred horse ranch for years and had a talent for seeing wolves traveling through a meadow miles away or bedded down in the distance. He was also skilled at telling individual wolves apart, probably from his experience identifying the many horses at his facility.

Around that time, Doug noticed a wolf behavior that I had never seen in Yellowstone: he saw a Druid yearling wade into the cold water of Soda Butte Creek and catch and eat two cutthroat trout. Coastal wolves in Alaska and British Columbia often fish for salmon, but fishing behavior is a rarity in Yellowstone where there are so many elk to hunt.

One day I watched as a Slough yearling buried an elk leg in a snowfield. It was clear that the young wolf did not fully understand the concept of hiding food for part of the leg stuck out of the snow. After that yearling walked off, another yearling came over, saw the leg protruding out of the snowfield, pulled it out, and carried it off. Caching meat for later use is a universal custom for wolves. I had often seen wolves carry off parts of caribou carcasses and bury them when I worked in Denali National Park in Alaska. Caching behavior suggests that wolves have a concept of the future. The digestive systems of wolves enable them to eat meat that would sicken a person, so they normally can consume old buried meat without a problem.

Back with the Druids, I saw 302 and a gray yearling traveling together one morning. They approached a bison bull that was bedded down. The yearling sniffed the bull's rear end, then poked him there twice with its nose. The huge bull, perhaps 1,800 pounds, stayed bedded, turned his head to look at the wolf, then kicked out with a hind leg. He missed, but the yearling, who was well under 100 pounds, got the hint and walked off.

On April 13, Light Gray was hanging out south of the Druid den area. Bright Bar, High Sides, and two gray females met up with him. The next morning, I got signals from 480 and 569 at the den forest and heard growls from there. I then saw wolves coming out of the trees: Light Gray, three of the young Druid females, and four yearlings. Soon there was howling from the den forest. The wolves out in the open howled back, then all of them, including Light Gray, went back into the trees. After that I heard more growling from the forest. Since I was not getting any signals from 302, the growling was probably from 480.

I spotted Light Gray bedded down east of the Druid den forest two days later. When he stood up, I could see that he was seriously injured. His fur was disheveled, a sign he had recently been in a fight. He moved off slowly and kept his tail tucked. After walking a short distance, he bedded down again, looking as though he could go no farther.

After seventy-five minutes, he got up and moved farther from the Druid den, his front paws shuffling forward only a few inches at a time. As he crossed a snow patch, his hind legs broke through the crust and he lay down, looking like he did not have the strength to go any farther. When he tried to stand later, he had a hard time getting up. There was blood on the snow where he had rested and blood on the side of his neck. He moved off a few yards, then had to lie down again.

I went west to look for other wolves, and when I got back five hours later, Light Gray was in the same area, walking sluggishly. He held his head low, nearly touching the snow. Then things got worse: I realized he was walking in circles, which indicated he had some type of neurological damage.

I checked on him four hours after that and found him lying on his side. Later he stood up, swayed back and forth, then stumbled forward. He sat with his head drooping down, then managed to lie down on his side. He was still moving slightly as I left for the night. When I got back early the next morning, he was dead. The temperature was 11 degrees Fahrenheit (–12 Celsius). He had probably died from a combination of his injuries and hypothermia.

Light Gray had been acting on a normal instinct for a male wolf. He apparently had wanted to visit females he had bred and help with their newborn pups. But 480 would have seen the situation differently. This was an outsider wolf invading his pack's den site and he could not tolerate that. 480 must have attacked him and beaten him up, then let him go. Whether he intended it or not, the bites he inflicted on Light Gray turned out to be fatal.

It was a tragic situation where both sides of the conflict, Light Gray and 480, were doing what was the right thing for them. If one of the young Druid females had run off with Light Gray, the story would have turned out much differently. But none of the females wanted to leave home, which set off the chain of events that led to the death of a wolf who was trying to care for females he had mated with and their newborn pups.

Erin Albers and I later walked up to examine the dead wolf. He had bite marks and blood on the right side of his neck and on his underside. There were also signs he was recovering from mange. We lifted him up gently and carried him down to a pickup. Erin drove him to park headquarters for a full necropsy, something that is often done when a park wolf dies near the road because the findings give us valuable information about the health of the population.

We figured that Dark Gray had also fathered some of the Druid pups that year. We occasionally saw him in Druid territory, but never near the Druid den site or any of the pack members. I wondered if that was because he had a better understanding of the risk of approaching the den or if he had less of a sense of responsibility than Light Gray.

I later thought about the mange on Light Gray. Perhaps 480 got an abnormal scent from that part of the wolf and instinctively knew it was something bad, something that could harm his family. When Light Gray entered the den forest, he unintentionally threatened the entire pack, including the new pups. I imagined 480 getting more and more aggressive to the young male and finally wounding him fatally.

AFTER THE DEATH of Light Gray, I went over to check on the Sloughs at the Natal Den and saw that 380, 526, and Hook no longer looked pregnant. 380 and 526 were based at the Natal Den and Hook was using the Sage Den downhill from the original site. A few days later I saw that those females were missing fur under their bellies, a sign all three were nursing. The other Slough wolves continued to hunt in Lamar Valley, west of the Druid den forest, which I thought could lead to more conflict between the two packs.

All the pregnant Druid females had their litters at their main den except for 571 who had chosen to have her pups at the pack's Cache Creek den site. In early May, 302 regurgitated twice to Dull Bar who also showed signs of nursing pups. I recalled the time when he had his fear of bull elk carcasses and stole regurgitated meat 480 had given to a mother wolf. Now 302 was the one feeding a nursing female.

On May 5, I saw Druid females 569, 571, White Line, and High Sides at a carcass. All four were missing fur under their bellies. A little while later, I saw that Bright Bar and Low Sides were also nursing. That meant the alpha female and all six two-year-old females had pups. In the following days, I got signals from 571 at the main den so she had apparently decided to move her pups in with the rest of the family.

One day I saw 302 approaching the road from the south. A car approached and 302 moved to another section of the road. He got up on the edge of the pavement, looked both ways, then crossed to the north and headed to the main den. Not many wolves fully understand the concept of cars moving in both directions on a road, but 302 did.

· I SAW MY first Slough pup of the year, a black one, at the Natal Den on May 11. It went in and out of the entrance several times, then followed a yearling downhill. 380 ran over, picked up the pup, and carried it back into the den. Later that day, mother wolf 526 fed on a fresh carcass south of the den area. She returned to the Natal Den and shared the elk meat with 380 by giving her two regurgitations, meaning it was one mother helping another mother. Then 526 found a pup at the entrance and carried it inside. After that, alpha male 590 arrived from the new kill and fed 380. Two days later, Laurie Lyman saw a yearling regurgitate to a pup at the den entrance. The young adults, who had been pups just a year earlier, were doing their part to keep the family's new litter well fed.

On the fifteenth, I observed a touching scene involving 380 and a pup. It was following her around and was often positioned under her belly as they traveled. At one point, it successfully climbed up and over a log. Mother and pup walked uphill together to the den. Then 380 bedded down and the pup climbed up on her face and went through her ears onto the mother wolf's back.

One day at Slough Creek, I saw an osprey flying by with a freshly caught trout in its talons. A bald eagle dived down and the osprey dropped the fish. The eagle swooped down and grabbed the fish off the ground as it flew past. The osprey tried to harass the eagle into dropping the fish, but it got away with the prize.

A grizzly came into the Slough den area in late May and 380 charged at it with her tail raised. When the bear saw the angry mother wolf, it ran off and other wolves joined the chase. 380 got behind the grizzly, lunged forward, and nipped its rear end. The wolves gave up the chase when they were satisfied the bear had run far enough from the den.

I had seen only one pup at the Slough den, the black one, and it seemed to be doing well. Other people saw it play with 526 for two hours so it was strong and healthy. Three females had looked pregnant and all of them seemed to be denning there, but there was just this one pup. In 2005 distemper had killed all but three of the sixteen Slough pups. The reason for such a low pup production this year did not seem to be distemper because that virus tends to kill pups later in their development. We never found out the reason for the poor denning season.

The Slough wolves continued to hunt in Druid territory in Lamar Valley, but always west of the den forest. 302 had a close call with them when he and other Druids were near the Yellowstone Institute and eleven Slough wolves appeared on top of the ridge to the west. The Sloughs raced along the ridge with tails raised as they followed a scent trail, either of 302 or of other Druids. I lost them when they ran downhill. Eight minutes later, 302 appeared on the ridge where the Sloughs had just been. He must have known the rival wolves were in the area for he ran east, but not all out, indicating he was not panicked. 302 reappeared farther east, calmly trotting toward the Druid den. He acted like he was confident he had outsmarted the Slough wolves.

I saw the lone Slough black pup nursing on 526, so she likely was its mother. One day it followed the adults when they left the den on a hunt, but it soon lagged behind. Two yearlings stayed with the pup, hovering over it, like they were concerned about its safety. Soon a mother grizzly with two cubs came into the area and the yearlings charged at them. The mother bear, wanting to protect her cubs, ran at the wolves. Meanwhile, the pup raced back toward the den. The yearlings distracted the bears and they never got near the pup.

ON JUNE 12, Carl Swoboda, who runs the Safari Yellowstone wildlife guiding company, saw three black pups at the Druid den. We soon saw a lot more pups, but it would take some time to get a full count. 302 spent a great deal of time at the den helping the mother wolves.

The Druids got a bison calf in mid-June, and 302 walked off from the site without feeding. 480 ate briefly, then also moved away. 569 and another mother wolf stayed at the site and continued to feed. The two males unselfishly let the females have preference at the carcass over themselves. This incident showed that 302 was doing much better at understanding that males have a responsibility to support females with young pups.

That day Calvin Johnston, a wolf watcher who was a farmer in Kansas, saw a young black Druid walk over to a nearby raven and grab it in its jaws. The wolf walked around with the motionless bird. A few minutes later, probably satisfied that the bird was dead, the wolf loosened its jaws and the raven immediately flew out of its mouth. Indigenous stories in Alaska and western Canada often have Raven as the main character, and he is referred to as the Trickster. This raven certainly tricked that wolf.

That summer the elk in Lamar Valley and at Slough Creek continued to feed in the low-elevation meadows longer than in previous years, and the Druid and Slough wolves got a lot of calves. Bob Landis told me he thought higher-than-average rainfall had caused the grass in the valley to be more lush than normal, which delayed the elks' departure to the high-elevation meadows.

The Lamar River was in flood stage and three elk calves got stranded on an island near the Druid den. The three mother elk could wade or swim across the river to nurse them, so the calves were doing fine. Then I got a report that a collared black Druid had approached the river in that area one evening. The mother elk ran over, cornered the wolf on top of the steep bank, and trampled him. He squirmed away and ended up at the bottom of the bank near the water. The elk surrounded him and trampled him once more. There were only two black collared wolves in the Druid pack: 480 and 302. I did a signal check early the next morning in that area and got a loud signal from 302. 480's signal was weak to the west, so it likely was 302 that got trampled. He was out of sight that day and the next, probably recuperating. I saw him three days after the trampling incident and he seemed all right. He was resilient to occupational injuries.

302 failed to get any of the elk calves stranded in the river, so he next tried to steal an elk calf from a grizzly. The grizzly was feeding on the calf, and 302 and two yearlings were close by. He and one of the young wolves rushed in and both grabbed an end of the calf. They ran off, jointly carrying it. When the bear gave chase, the two wolves had to drop their prize. The bear reclaimed the carcass, but 302 was not ready to give up yet. As the bear continued to feed, 302 approached and the bear promptly chased him. That enabled

the yearlings to run in and get away with pieces of meat. 302 went to one of the young wolves, watched as it fed, then walked off, letting the yearling eat all the stolen meat. After the bear left the area, 302 found a calf leg and fed on it. I had to admit that I was impressed by 302. He was taking on risks to get food for his pack and treating lower-ranking wolves with respect. 302 had become a different wolf from his earlier slacker days.

20

Occupational Injuries

AFTER 302'S RUN-IN with the bear, I saw eight of the Druid adults chase a bull elk and easily catch up with him. Some of the wolves nipped at his hindquarters but did little damage. The bull ran to the river, jumped down the bank, and waded out into shallow water. Getting behind the bull, 480 grabbed the hindquarters. The bull kicked back and bucked up and down, shaking the wolf loose. 480 grabbed his rear end once more and got kicked off. The bull charged the wolves and kicked back at them once again. 480 got another bite on the rear of the elk. The other wolves ran in and bit him there and on the sides and back. They pulled him down. He got up, shook them off, and ran into deeper water, forcing the wolves to swim after him. They encircled him, but now that they had little leverage, they gave up and dog-paddled to shore. The bull was just too strong for them.

After that incident, I heard that a young black Druid was seen chasing a cow elk. He grabbed a hind leg and stubbornly held on as she kicked at him with her other hoof. One kick sent the wolf flying through the air.

As I read back over my field notes, I was struck by the number of cases where Yellowstone wolves were injured or killed while hunting elk and bison. Those incidents made clear how dangerous it is for wolves, who average about 100 pounds as adults, to fight with adult elk that can weigh from 300 pounds to 700 hundred pounds and bison that can get up to 2,000 pounds. The following list of hunting injuries inflicted on wolves is based on my own sightings and experiences in Yellowstone and reports from other people.

In *The Rise of Wolf 8*, I wrote how a Crystal Creek alpha male and a big Rose Creek male both died after being stabbed by a bull elk's sharp antler point. I saw wolf 8 fight an elk in Slough Creek and get kicked in the head and stomped on his back. A younger wolf who was helping 8 was behind the elk and twice got kicked with a hind leg. On the last day of his life, 8 got in a fight with another elk in the same creek and drowned in shallow water there, probably after being kicked in the head and knocked unconscious. An examination of 8's skull found extensive damage caused by kicks from large prey animals. Two of his four canine teeth were missing and a third one was broken off. Several other teeth were missing or broken. Jim Halfpenny measured the distance between two broken teeth on the right side of his jaw and found that it was about the same width as an elk hoof. Wolf 8 had been kicked in the face so hard that the blow broke off two of his teeth.

In *The Reign of Wolf 21*, I wrote how I saw a bison kick back at a pursuing wolf and hit it with such force that the wolf flew backward in a complete circle before crashing to the ground. In another incident, a male wolf was found dead at the base of a cliff. It looked like he had been butted or kicked off the top by an elk and died when he landed.

Doug Smith saw the Mollie's wolves have an all-out fight with a bison. During the battle, one wolf was kicked and thrown fifteen feet through the air. Another was hooked by a horn and tossed a similar distance. A third wolf broke her leg in the fight and later died from additional injuries.

When I was watching the Druids in 2000, I saw wolf 21's mate, 42, get kicked by a cow elk and knocked down, but she got right up and continued the chase. A group of seven pups chased a bull elk and nipped at his hind legs and rear end. A pup got kicked in the head or chest, flew backward, crashed to the ground, then jumped up and continued the chase. A similar thing happened to one of the yearlings when it was chasing a bull elk. Druid beta male 253 got stomped by a cow elk's front hooves. I watched as another female Druid chased a bull elk that kicked back and knocked her down, but she stood up and resumed the pursuit.

Earlier in this book, I described how 253 was kicked by a bull elk and sent sailing down a riverbank. He landed hard, rested for six hours, then recovered and walked off. In another instance, when a yearling bit a bull elk on the nose, the bull lifted the wolf off the ground and tossed him aside. Another yearling bit into the bull's shoulder, and he also threw her off. Two yearlings chased a lone bison and were both kicked with such force that they flew backward through the air.

An alpha female was found dead close to a bull elk carcass. He had apparently kicked or gored her and the wolf's family later finished him off. Another female was kicked in the throat and died from the blow.

In 2005 Mollie's male 379, the wolf who attacked 302 and then was defeated by 480, died. When Dan Stahler did the necropsy, he found that the wolf had three broken ribs and bruising on the skin under his hide. There were lacerations on his lungs and liver, which had caused severe internal bleeding. All those injuries indicated that he had been kicked by a large prey animal, either an elk or a bison.

The Leopold alpha female broke a hind leg, probably when she was kicked by an elk, but the bone fused on its own. I once saw a young Slough male survive a devastating kick. He approached a big bull elk standing on a knoll and lightly touched a hind leg with his nose, probably to test the bull's reaction. The elk kicked back and hit the wolf so hard he fell fifteen feet down a steep section of the hill. He was not seriously injured because he landed in soft snow. The yearling ran right back up to the bull but this time got face to face with him. Five other wolves joined him, but they were too intimidated by the aggressive bull and soon left.

A Slough alpha male confronted a cow elk in a creek. She reared up and stomped on his back with both front hooves, then did the same thing to the pack's alpha female. When the male swam after her, the cow turned to face him and stomped him once more.

After the events described in this book, I saw a wolf get headbutted by a bison and tossed upwards. He had the bad luck to land on the cow's head, and she flicked him up once more. He landed hard that time. I also saw an old alpha male engage in an epic battle in the Lamar River with a cow elk who was much larger and stronger than he was. He eventually leaped up, grabbed her throat, and held on for several minutes until she collapsed and died. We later learned that the wolf's lower jaw had been broken months earlier. Like wolf 8, he must have been kicked in the face by an elk. That old break would have caused him unimaginable pain when he had to use all the force in his jaws to finish off the elk.

Mike Phillips, the original lead biologist of the Yellowstone Wolf Project, worked in Alaska early in his career and did a study where he examined 225 wolf skulls. Twenty-five percent of them showed evidence of blunt force trauma such as broken jaws or damaged skulls, most likely inflicted by kicks from moose. An earlier study of 2,134 Alaskan wolf skulls by H. Haugen found that 36 percent had injuries caused by prey animals. In the book *Wolves on the Hunt* there is a photo of a wolf skull with a hole on the top made by a whitetail deer hoof. That injury is proof of the force behind a hoofed animal's kick. Rolf Peterson, who has examined dead wolves at Isle Royale National Park for decades, told me that about half of those wolves had one or more broken ribs caused by run-ins with moose.

Sue Ware, a paleopathologist at the Denver Museum of Natural History, examined many skeletons of Yellowstone wolves including wolf 8's. After describing his accumulated injuries throughout his long life she concluded, "I do not understand how an animal could live through this."

It is hard for us to comprehend how dangerous it is for wolves to engage in face-to-face combat with large prey animals. Think about having to fight an opponent that could be three to twenty times your size. Then try to imagine dodging four rock-hard hooves that could strike you in the head or trample you. If your opponent is a bull elk or bison, you could also be pierced in the chest or belly by a pair of antlers or horns. That means bulls can damage you with six different weapons while you have only one: your jaws.

Any wolf must accept the possibility that whenever it goes out on a hunt it could get kicked in the head, stomped, gored, or trampled by an elk. With bison those risks are the same, but it could also be thrown through the air by a backward kick or headbutt from a 2,000-pound opponent. It must be willing to repeatedly go into combat, take on the risk of those potentially fatal or crippling hits, and absorb whatever injuries it receives so the wolf can fulfill its responsibility to feed its family.

I consulted with Dan Stahler and Kira Cassidy, and they told me that the Wolf Project found that about 13 percent of all known wolf deaths in Yellowstone were attributed to attacks by prey animals.

But that is not the only area in which wolves suffer violent deaths. Wolf Project records reveal that 50 percent of known mortality is due to fighting among wolves, usually over territory. That means that about 63 percent of Yellowstone wolf deaths result from wolves being fatally wounded during hunts or in fights with neighboring packs. There was no reason to think that 302 would be exempt from those potential fates.

I often wondered how these fatality rates would compare with those of early humans who lived in small bands, hunted

large animals with primitive weapons, and often had to battle
rival groups who wanted to take over their territory. I suspect
that their causes of death might have been similar to those
of the wolves.

Dan MacNulty, a former Wolf Project staff member
and current professor at Utah State University, was the
lead author of a 2011 research paper on the success rates of
wolves hunting elk in Yellowstone. He and his coauthors
found it ranged from 1 percent to 10 percent. Rolf told me
that in Isle Royale National Park the wolf success rate when
hunting moose is about 5 percent. In Denali National Park,
Gordon Haber found that when wolves hunt Dall sheep they
get them only 3 percent of the time. I have not tried to tabu-
late the success rates of elk hunts I have seen in Yellowstone
but would guess that it would be very low, close to 1 percent
or 2 percent.

After writing about the threat of severe injuries and death
when packs hunt, I thought about how my years of watching
wolves have taught me they are born to hunt and enthusias-
tically take on whatever risks that occupation entails despite
the dangers. I recalled a day in March of 2009 when Bob
Landis and I witnessed an encounter between the Druid
wolves and a bull elk.

The wolves chased a group of bulls and singled out one
of them. The bull ran to the top of a high ridge, then turned
to face the wolves. He stood his ground and kicked out and
down with his front hooves at the nearest wolf, like a wild
stallion. As he charged at the wolves, another wolf darted in
and grabbed a hind leg. The rest of the pack rushed in and
attacked the bull. They pulled him down, but he was right

at the edge of a steep drop-off. As he struggled to get up, the bull slipped over the edge and fell down the slope. The Druids fearlessly ran down the nearly vertical snowfield as the elk slid downhill, at times on his back. The wolves maintained their balance with the grace of champion Olympic skiers as they pursued him. The bull slid to a stop in a less vertical area, and the wolves were on him before he could get up and quickly finished him off. That was one hunt where the Druids risked their lives and got away with it.

21

A New
Pack Forms

A YEARLING BISON DIED south of the road near the
Yellowstone Institute on July 2. The next morning, a
big herd of bison were gathered at the site, sniffing
the dead animal. Local people call that a bison funeral. We
do not know if the animals come together because they knew
the one that died or whether they are just curious. I have
sometimes seen bison lick the face of the dead animal and
felt in those cases they did know it.

A couple of days later, I went up on Dead Puppy Hill, a
vantage point I often used to monitor the Druid den across
the road to the north. I counted four mule deer and five elk
near the den site. I had frequently seen prey animals walking
around the Druid and Slough dens, seemingly unconcerned
about the presence of wolves. Usually the wolves ignored
them, perhaps assuming only strong and healthy prey

animals would display such confidence, animals the wolves knew they could not take down.

Since the Druids' main hunting area was south of the road and the den was to the north, the adult wolves had to make frequent crossings. That often caused problems. A typical situation occurred when 302 and 569 appeared south of the road. They wanted to get back to the den, but four cars were now stopped at the place where they hoped to cross. I was wearing my ranger uniform, so I went down and spoke to the drivers, who agreed to move on.

I later received word that an adult wolf and three gray pups had crossed the road to the south. Then two of the pups panicked and ran back toward the road. Drivers stopped and, although they meant the pups no harm, their presence blocked the young wolves from getting back to the den. Campground host Ray Rathmell heard about the problem, went to the crossing point, and got the cars to move on. As he drove off, he looked in his rearview mirror and saw the two pups crossing to the north. I later saw the third gray pup south of Soda Butte Creek in an area where adult wolves had also been spotted.

The next day, we saw nine black pups and eight gray pups up at the den. Including the gray pup that had crossed to the south, the Druids had eighteen pups. That was close to the record count of twenty-one back in the spring of 2000 when 21 and 42 were the alpha pair.

After that 302, 569, and two black adults crossed the road to the south with twelve pups following them. The pups seemed to be from different litters for some were big and others small and medium sized. Based on the mating activity we had seen, the pups would be ten to thirteen weeks old. The adults crossed the creek, but the pups hesitated when they reached the water.

The pups looked like they were going to run back across the road, so I drove partway toward the crossing point and blocked traffic. I could see pups and Dull Bar, one of the mothers, between the creek and the road to the south. One brave pup swam all the way across the creek. Another attempted to cross but turned back midstream. It looked like a drowned rat when it came out of the water. Eventually three of the pups made it across the creek. Other pups and Dull Bar were now trying to return to the den.

302 appeared on the south side of the creek and immediately took charge. He swam across the creek and got the stranded pups back across the road to the north. As he escorted them to the den, he stopped frequently, looked to the south, and howled, probably trying to get other pups that might be trapped between the road and the creek to come to him. We lost him and his group of pups as he continued north toward the den. The three pups that got across the creek were out of sight in the trees by that time. I had the thought that the stranded pups who were helped by 302 must have seen him as their hero.

The next day, July 10, I heard that Bright Bar had led thirteen pups to the creek. They all crossed the water successfully and went into the forest east of Dead Puppy Hill. That left just two pups at the den. Two days later, I did not get any Druid signals in the area and assumed the adults had now moved all the pups across the creek. A flight in late July saw the Druid adults and at least ten pups near Cache Creek.

We got a mortality signal from male Slough wolf 629 in Lamar Valley, several miles west of the Druid den. He was the male Druid wolf who had successfully joined the Slough pack in 2007 and worked his way up to the second-ranking male position. A crew went out and found him. There were no bite marks or other injuries on his body. Part of his remains were brought back, and genetic testing showed that 480 and 569 were his parents. Another test found that he had died of distemper.

Emily Almberg had started a PhD program at Pennsylvania State University and was studying diseases in wild wolves. She tested three other Yellowstone wolves that had died in 2008 and found that the cause was also distemper. This was the virus that had killed most of the Slough pups in 2005. Emily told me that if an adult wolf had never been exposed to distemper, it would have no immunity and could easily die. I later talked with famed wolf biologist Dave Mech and he told me more about how distemper is spread through wolves via direct contact, such as saliva. If one wolf chewed on a bone and another wolf gnawed on it soon after, it might pick up the virus. The virus would not survive in a den from one spring to the next.

As of the beginning of August, we had not seen the black Slough pup for two weeks. Doug Smith sent a crew to the Slough den site and they found its remains. That meant the Sloughs had no surviving pups that year.

The Uinta ground squirrels seemed to be at a much higher density than normal that summer, and wolves chased and ate them far more frequently than I had seen in the past. The squirrels were fat as they prepared for hibernation, so they made a good meal for a wolf. A big one might weigh ten ounces. Even Slough alpha male 590 ran around after one and got it. When I worked in Denali National Park in Alaska, I saw wolves and grizzlies catch and eat ground squirrels much more often than in Yellowstone. Those Arctic squirrels can get up to two pounds, so it is more worthwhile for large predators to go after them. Indigenous Peoples there used to make their winter parkas out of ground squirrel fur and called the animals "parky squirrels."

WITH THE DRUIDS spending most of their time up the Lamar River at Cache Creek, the Slough wolves were free to hunt in Lamar Valley. I saw them make good use of that opportunity one day when they chased a bull elk into the river. They swam after him as he waded away on his long legs. After a while, he tired out and the wolves were able to surround him. When he tried to charge at wolves in front of him, two others bit him on the hip and shoulder. He bucked up and down, but they held on. Then 590 leaped up and grabbed the bull's throat. After some struggling, the wolves pulled him down into the water. Losing the last of his strength, the bull allowed his head to sink under the surface and he drowned. The chase and fight in the river took about twenty minutes.

By mid-August, the annual bison rut was in full swing and the bulls were getting into violent fights. One was gored to death in Little America. The Slough wolves picked up the scent and fed on his carcass. Soon after that, another bull was seriously injured in a fight with a rival in Lamar Valley and had to be shot by the rangers. 302 and three other Druids came back to the valley not long after that, found a different dead bison bull in the Chalcedony rendezvous site, and fed on him.

When 302 and those wolves returned to Cache Creek and fed the pups with regurgitated bison meat, it probably tipped off the other Druid adults to the new carcass. I later spotted a group of Druids, including 480 and 569, heading to Lamar Valley from Cache Creek on the reverse route of 302's party. I counted twenty-four incoming Druids, including seventeen pups, one short of that year's high count of eighteen. Three other pack members were already at the bison carcass. The two groups got together, and the twenty-seven wolves had a big greeting ceremony. 569 caught a plump squirrel and a pup ran over begging for it, but 569 kept it for herself.

I then spotted seven Slough wolves traveling east toward the bison carcass. The Druids must have gotten their scent, for they had a big group howl. The Sloughs stopped, looked in the Druids' direction, then ran west, away from the much larger pack. The days when the Slough wolves greatly out-numbered the Druids were long gone—what had been the smaller pack was now too big for the Sloughs to push around.

The next morning, a mother grizzly with one cub approached 302 and a group of pups at the Chalcedony rendez-vous site. 302 positioned himself between the pups and the bears. When the mother grizzly saw the big wolf, she veered away. Then 302 began to stalk the bears. The pups followed and the wolves were soon arranged in a circle around the bear family. 480 and four other adults saw what was happen-ing and raced over to help 302 guard the pups. The grizzly sow charged and swatted at the wolves. One wolf nipped the bear cub on the rear end. Then 569 and more adults ran in, and the pack drove the bears off. After that the Druids had a big rally where they jumped up on each other, like a football team after winning a championship game. The pups had stuck close to 302 throughout the encounter. I also noticed that they liked to bed down close to him, which indicated they felt secure being with him.

That day I counted twenty-nine wolves and realized that four adults were missing. If they reunited with the rest of the wolves, the pack would number thirty-three, just five short of the all-time high count of thirty-eight Druids in 2001. Thinking about that period in 2001, I realized that the relationship between 480 and 302 reminded me of the partnership between 21 and his son 253. Like 253, wolf 302 was now contributing a lot to the pack, which made it easier for 480 to fulfill his alpha male responsibilities of feeding and protecting the family as well as raising the pups.

ONE EVENING A mother grizzly and her two first-year cubs came into the rendezvous site to feed on the bison carcass. Several young adult wolves charged and the bear family ran off. One cub panicked, left its mother, and raced toward the nearest tree. The wolves caught up and nipped at it. The cub fought back vigorously and broke away. A single wolf ran alongside it and bit the cub. It stopped and stood up on its hind legs to confront the wolf.

More wolves ran in and the cub continued to defend itself as it ran toward the tree. At one point, it stopped and snapped at a group of wolves surrounding it. Finally the young bear broke through the circle of wolves, got to a big aspen tree, and climbed up. When it made it four feet up the trunk, a wolf leaped up, bit it on the rear end, and yanked it to the ground. The cub climbed up the tree again but fell off when it got up ten feet. It tried a third time and fell once more. The mother bear and other cub were nearby, so this time the lone cub ran to them. The three bears had a standoff with a group of Druids, then the wolves lost interest and left. I had the sense that the incident was a game to them, not a serious attempt to kill the cub.

While the other Druids were taunting the mother and her cubs, 302 and a big grizzly were trotting toward the bison carcass, twenty yards apart. The wolf got to the carcass first and started to feed, ignoring two huge bison bulls that were moving toward him and the grizzly. The bulls went for the biggest target, the bear, and it moved off. Meanwhile 302 occasionally glanced at the bison as he gulped down meat as quickly as he could. When the bulls switched their attention to 302, he grabbed another couple of bites before slipping away just in time to avoid being trampled or gored. I was impressed at how well 302 judged the situation. He seemed to guess, correctly, that the bison would go after the bear first, then come after him. That enabled him to get a good meal before running off. When I was a kid, I watched Yogi Bear cartoons, and he was said to be smarter than the average bear. That episode with the grizzly and bison caused me to think that 302 was smarter than the average wolf.

We stopped seeing ground squirrels on August 21. They tended to go underground for their winter hibernation as soon as they had enough fat on their bodies. Staying active longer only gave predators more time to catch and eat them. I saw the first aspen leaves turning yellow four days later, another sign of the coming winter at this high elevation. In late August, I reached my three thousandth day in a row of getting up early to go out to study wolves—over eight years. My goal now was to reach the ten-year mark. I had seen wolves on 98.7 percent of those three thousand days.

ON SEPTEMBER 3, I saw 302 with two younger adults and seven pups at the rendezvous site. The alpha pair and nine adults were west of there. A few miles farther west, six Slough adults were near a cow elk they had chased into the Lamar River. They were once again trespassing into Druid territory, and it looked like this time there was going to be a battle between the two groups.

The Druids must have spotted the Sloughs for they ran that way. The intruding wolves scattered. One black ran into the river, near the elk. The Druids raced over, splashed through the water, and attacked the outsider. Then 569 and five wolves spotted another Slough wolf, left the one they had caught, and chased their new target. Other Druids stayed behind and continued to bite at the wolf in the water. They soon ran off, looking for other Sloughs to chase. Some of the Druids then returned to the river and pulled the Slough wolf onto the bank. It was not moving and must have died from the bites and blood loss. I saw a group of nine Slough wolves to the west, including the alpha pair. They howled, giving away their location. The Druids charged in their direction, and the Sloughs fled west toward their territory. That satisfied the Druids, who ended the chase. Neither 480 nor 302 had been involved in the fatal attack.

It turned out the dead wolf was female 526. The next morning, Erin Albers, Kira Cassidy, and I walked out to the river and waded to the other side. The wolf was still partly in the water. There were bite marks on her body and one puncture wound on her face. 526 had repeatedly trespassed into Druid territory and eventually her boldness cost the wolf her life.

I reviewed the twelve incidents in this book where the Druids caught wolves from other packs in their territory and found that in nine cases the wolf was let go or allowed to escape. That included two incidents in 2004 where the Druids caught Slough wolves and spared their lives. But in 2005, the Sloughs killed Druid yearling female 375 in Druid territory. In 2007 and 2008, the Druids caught and killed three Sloughs.

I realized the incidents had a pattern. Prior to the Druid female's death, the Druids had spared the life of two wolves from the Slough pack, but they killed three Sloughs after the Sloughs first killed a Druid wolf. The Druids had a valid grudge against the Sloughs and might be said to be operating under the ancient rule of retaliation described in the Bible as "an eye for an eye, and a tooth for a tooth." In contrast, the Druids killed no wolves from other packs other than Light Gray.

I thought back to his story. The Druids caught him twice when he came into Druid territory. In the first case, 480 beat him up and let him go without any obvious serious injuries. In the second incident, after he repeatedly intruded into their den area, they wounded him then let him run off, but he later died of his injuries. I think in that case they were trying to drive him away from their den site and did not intend to kill him.

Three days after 526 was killed in the river, we got a mortality signal from Slough alpha female 380 toward the west end of Lamar Valley. Later a Wolf Project crew went to the area and found her body. It looked like two or three wolves had attacked her. The killing bite was to her throat. The Druids were miles to the east that morning as they had been the previous day. Kira had gotten some signals earlier that day and felt that the Slough wolves had been up on Specimen Ridge. That put them in Agate territory, which made them the prime suspects in 380's death.

The Slough pack was in a period of contraction. There were no surviving pups that year and the number of adults had dropped from fifteen in early May to eleven. The pack's new alpha female was the young uncollared black known as Hook.

All sixteen Druid adults were still in their pack, so they significantly outnumbered the Sloughs. The high count of Druid pups had been eighteen, but by late September we were seeing only five. As in past years, we never knew what happened to the missing pups. We looked for identifying marks that could help us tell the yearling Druids apart. Of the two big black males, one had a large white blaze on his chest and the other had a smaller one. We called them Big Blaze and Small Blaze.

On October 7, I saw most of the Druids west of the Chalcedony rendezvous site. 569 and other adults were playing chasing games with the pups. 302 came over and romped around with the pups, much like I had seen 21 do. 302 was an old wolf by then, but the way he joyfully played with the pups made him seem more like a yearling.

Later I spotted five of the younger Druids at Slough Creek: four male yearlings and their older sister Dull Bar. Two black males did raised-leg urinations. It was a bold move for the group to trespass into the center of the Slough territory, but they got away with it. The group stayed in that area for the next three days.

On the tenth, 590 and six Slough wolves were back in that area, and they got the scent of the Druid intruders. Right after that, I spotted the five Druids charging at the Slough wolves with raised tails. 590 and his group countercharged but slowed down when they got close to the Druids even though they outnumbered the opposition. Then 590 ran off, and the wolves with him scattered. The Druids chased some of them, but they mostly stayed in close formation while the Sloughs were disorganized. I noticed that the Druid group controlled the high ground, which enabled them to see where most of the individual Slough wolves were. During the melee, Jan Taylor, one of the wolf watchers, saw three black Druids grab alpha male 590 and pin him. He escaped and ran off with the Druids chasing him. 590 was a big tough-looking male, but he could not cope with these much younger Druids.

This defeat of the Sloughs by a smaller force of Druids reminded me of how 480 had won the battle against a larger group of Mollie's wolves with just a few pups to support him. These young Druids must have learned something about tactics from him. The four male yearlings were likely his sons, so they also inherited his physical traits. Their group went back to Lamar Valley the next morning and rejoined their family.

The main Druid pack, including the alphas and 302, went into the Slough territory soon after that. I wondered if the yearling males had headed that way and the others had followed. The Slough wolves wisely stayed out of their way. Most of the Druids returned to Lamar Valley the next morning, but all five male yearlings stayed behind.

On October 24, we found 302 with those yearlings west of Slough Creek. They were traveling with a black female Agate wolf who was flirting with them. By late October, 302's subgroup was back with the Druids and the black female had returned to the Agates.

The five male yearlings frequently traveled west into Slough territory in early November. When they howled one morning, a nearby larger group of Slough wolves, including alpha male 590, ran away from them. At times the rest of the Druids joined the yearlings as they traveled to Hellroaring Creek and areas farther west without resistance from local packs.

I had not seen this behavior before. Normally young adult wolves leave their families and seek out a mate and a territory by themselves, rather than disperse as a group. I figured that this band of brothers would find some females, recruit them to their group, then stake out a territory as their own. The expeditionary group usually explored in a westerly direction.

WE GOT A report of a Leopold wolf that had such a bad case of mange that a law enforcement ranger saw it fall down on the park road. Mange was a significant threat to Yellowstone's wolf packs because wolves need thick fur to insulate them from extreme cold weather. Thermal images of wolves with mange indicate that the animals need to double their energy expenditure to keep themselves warm in winter. A patchy fur coat would be like a person wearing a down jacket with big holes in it.

The ranger was going to shoot the wolf for humane reasons, but the wolf got up and walked away. Soon after that, another Leopold male was found dead. He was very thin and also had mange. There were no wounds or injuries on the wolf, so the mange must have severely debilitated him. The Leopold pack was going through a tough year. The combination of conflicts with neighboring packs, an outbreak of distemper, and now mange had wiped out many family members. By the end of the year, the pack no longer existed. A vacant wolf territory is usually soon colonized by a newly formed pack, so we monitored the area to see who would claim it.

On November 16, we saw 302 and the roving gang of Druid males near Tower Junction with a gray female from the Agate pack. Two days after that, the Agate Winter Study crew spotted 302 in the Agate pack's territory at Antelope Creek, about five miles south of Tower Junction. He was with a group of nine wolves that included his five nephews, Agate female 642, and other females from her family. The Druids and the Agates were doing a lot of interacting—exactly what you would expect from a group of males meeting females from another pack.

The Druid crew and I went up and joined the Agate crew. We spotted 302, the Druid yearlings, and five Agate females: 642, a black adult, a gray adult, the gray pup who had been run over by the car in January and lost half her tail, and the 06 Female. 06 had blood on her face, likely because she had recently been beaten up by her aggressive sister 693, the other gray Agate adult in the group.

Later that day, I saw fourteen other Druids near the Slough den. Even without 302 and the young males, the Druids outnumbered the Slough wolves, so they could go anywhere they wanted in the rival pack's territory. There had been a lengthy period when the Sloughs pushed the smaller Druid pack around, forcing them to retreat to the far reaches of their homeland. Now the Druids were the superior pack and could make incursions into the Slough territory without being challenged.

302's group of Druids and Agate females was on a bison carcass south of Tower Junction on November 19. The following day, Dan Stahler saw 302 and nine wolves west of Hellroaring Creek. Another spotter saw a gray Agate female do a double scent mark with 302, which indicated they were the alpha pair of the new group.

I got a good look at those nine wolves with 302 on the twenty-second. The Druid members were 302 and the five male yearlings: two blacks, Big Blaze and Small Blaze, and three grays. One of the grays had brownish sides so we called him Big Brown. Another one was called Medium Gray. The third gray had no distinct markings and remained unnamed.

The four females included the black who was probably the first Agate to be with the males and the gray who seemed to be the alpha female. Three of these females (642 and two soon to be known as 692 and 693) were born to former Agate alpha female 472, who was a daughter to 21 and 40. That meant that 302 was starting his new pack with three of 21's granddaughters. The 06 Female, another one of 21's granddaughters, was the female who had dropped out of the group earlier. Soon it was clear that 693 was the top female. She had a history of abusing her sister, so that was likely a factor in 06's departure from the group. For now 302 was an alpha male in a pack of ten wolves.

Would he succeed in fulfilling the heavy responsibilities of that position or revert to some of his earlier less-than-heroic behavior and let his new group down?

The Winter Study crews came up with a temporary name for 302's pack: the Dragates. It was a combination of Druid and Agate. If they found a territory of their own and settled there, we would name them after the area. The alpha pair, 302 and 693, did a lot of interacting, and the five younger males flirted with the three other Agate females.

That winter we noticed that wolves were squabbling among themselves at carcasses more frequently than normal. Dan told me he thought the weakest elk had died in the previous harsh winter, while the strongest ones survived. They fed well in the spring and summer, and now the vast majority were robust and healthy and hard for the wolves to catch and kill. Dan felt the primary issue facing the wolves was not fewer elk, but a lack of vulnerable elk. That caused more fighting between packs as they strayed into neighboring territories to look for prey. We also saw quarreling among pack members at carcasses. That kind of dissension within a wolf pack often leads to lower-ranking adults setting off on their own to try to start their own packs. In this case, it was a group of male yearlings who left, along with the much older second-ranking male in the pack.

In late November, 302's group was bedded down in the Leopold pack's old territory at Blacktail Plateau. 302 had been born there in 2000, lived there for four years, and would still know the area well. The Leopold pack had fallen apart by then, so it looked like the new group was claiming that vacant territory. We decided to officially call them the Blacktail pack. 302 had moved back home.

In mid-December, I was getting good signals from 302's group but could not find them. I tried a trick that sometimes worked: I followed a raven and it took me right to the wolves to the west of where I had been searching. I was eager to find the pack because breeding season was approaching. I noticed that the males and females were flirting a lot more than earlier. Big Brown was trying hard to get 693's attention, but she was more interested in 302. He still had whatever it was that drew females to him.

PART 6

2009

22

The 06 Female

A T THE START of 2009, the Druids numbered thirteen: 480, 569, six younger adult females, and five pups. 302's Blacktail pack had settled at eight members: 302 and 693 (the alpha pair), females 642 and 692, and four male yearlings (Big Brown, Medium Gray, Big Blaze, and Small Blaze). The fifth Druid yearling, an uncollared gray, had apparently departed the group.

There were only seven wolves in the Slough pack because alpha male 590 had left. Dan Stahler did a flight in mid-November and saw him alone out of the park, north of Gardiner, Montana. It was elk-hunting season in that area and 590 was feeding on gut piles where hunters had field dressed elk. A young black male in the Slough pack now seemed to be the new alpha. He may have challenged 590 and displaced him.

The Agate pack had also lost members. Former alpha male 383 had dispersed. Agate alpha female 472 was his aunt and the younger females in the pack were likely 383's sisters or daughters, and he needed to be with unrelated females. When I saw the Agates in early February, they were down to four: alpha female 472, a black male, and two gray females. The two gray females were the 06 Female, who had returned after temporarily staying with 302's group, and her younger sister, the 07 Female. I saw 06 flirt with the black male, so he was probably unrelated to her.

I watched a video of the Blacktail wolves around that time. 302 was guarding 693 when two of the younger males came over. 302 raised his tail to them, signaling his higher rank. The younger wolves submissively rolled on their backs, acknowledging 302's alpha status. Big Brown was the beta male, and 302 spent more time dominating him than the other young males.

Later in the footage, all four yearling males gave 302 submissive greetings. 302 stayed close to 693 that whole time. At times the younger males tried to approach 693 but would back off if 302 stared at them. 693 often stood in front of 302's face and averted her tail, indicating that she was getting close to being in season. When she walked off, he followed her closely, trailed by the other four males in single file. She looked like a queen with her retinue.

On February 10, I saw 06 avert her tail to the Druid gray yearling who had temporarily left the Blacktail pack. A few days later, 302, 693, and three yearling males were in Little America. Nearby we saw 06 get in a mating tie with the gray male she had met up with earlier. Five hours after he bred 06, the same gray male mated with her mother, Agate alpha female 472. When the black Agate male approached, the gray chased him off. That black male reappeared on Valentine's

Day, found 06 alone, and they bred. That was the second male she mated with.

In mid-February, 06 was seen feeding on the fresh carcass of an elk she had killed by herself. The gray male she had mated with had rejoined the nearby Blacktail wolves. Soon after 06 finished feeding, she met up with the gray again. This time he was with his brother Big Blaze, and she mated with him twice, her third partner. Big Blaze mated with 472 later in the day. That meant that two of the Blacktail males had mated with both 06 and her mother.

In the early morning of the eighteenth, 06 was near Tower Junction with four of the Blacktail males: 302, Big Brown, Medium Gray, and Big Blaze. 06's sister, the Black-tail alpha female 693, was bedded down nearby. The Blacktail males were ignoring her and giving all their attention to 06 as they followed her around, mostly in single file. Their behavior indicated that 06's pheromones were advertising her readiness to breed, while 693 had apparently gone out of season. She was old news to the males and her retinue was now fixated on 06, the sister who 693 had so often bullied and beaten up.

Medium Gray was especially persistent, which annoyed 06. She repeatedly lunged at him and bit him on the head and tail. I have often seen female wolves aggressively reject suitors. The males always seem eager to mate with any female coming into heat, but the females decide who mates with them and who does not.

693 finally joined her sister and the males, who by this time were bedded down. 693 stood over 06 in a dominant posture with a raised tail, then walked off and bedded down.

I think 693 knew she could not control the males and had given up on keeping them from her sister. I saw that 06 had an old scar below an eye, likely caused by a bite from 693 during an earlier time. To me it looked like 06 was exacting revenge on her sister for years of mistreatment by outcompeting 693 for the attentions of the Blacktail males.

After 693 gave up and walked off, Big Blaze went to 06, but she snapped at him just as she had snapped at Medium Gray. 302 came over and snarled at Big Blaze. When the young male approached her again, 06 lunged and bit him on the side of his neck. He did not fight back and seemed to be just plain afraid of her. We had noticed that during a previous altercation she had yanked out a patch of fur from near his eye. 06 was not a female to be messed with.

Big Brown joined the two of them and put a paw on 06's hip. Her mood had not improved, and she bit him, too. When he pestered her again, she grabbed him by the neck. That could have been a killing bite, but apparently it was meant as a warning because she let him go. I noticed that Big Brown had blood on his jaw that must have been from the bites 06 had inflicted on him. None of the yearlings retaliated when she bit them. Everything I witnessed that day confirmed what I had seen through twenty-one mating seasons in Yellowstone: male wolves accept rejection, do not bite back when bitten, and do not force females.

302's reactions to all this attempted mating activity impressed me. He acted like a chaperone, guarding 06 from the unwanted advances of the young males, and did not pester her like the others. I saw 302 gently lick 06's back and lightly rest his head on her when they bedded down next to each other. She accepted his attentions and did not snap at him or bite him. Based on his many years of romancing females, 302 seemed to know how to treat them. In human terms, he was suave. I now understood why the females liked him so much—he treated them the way they liked to be treated.

When 06 walked off, all the males followed her, once again completely ignoring 693 who was bedded down nearby. 302 joined 06 and calmly sniffed her. She averted her tail to him in response, something she had not done to the younger males. After that 06 pressed her side against 302. He put his chin over her back again, a gentle courting move. 06 continued on and all the males followed. The group went into some trees and I got only brief glimpses of them. Nineteen minutes later, I picked them up again. Medium Gray was in a tie with 06. He was the fourth male to mate with her.

People later lent me videos they shot when 06 was with 302's group, and I studied details I had not seen. They showed 302 pinning the younger males and getting between them and 06, like he was protecting her from them. At times 302 bared his teeth at the males and growled at them. He often walked away to give her some peace, but he came back when the other males pestered her. Once when Medium Gray mounted 06, 302 knocked him off and bit him. Later he stepped away, then rushed back when he saw Medium Gray mounting her again, but Medium Gray was already in a tie with her.

I recalled seeing 302 standing apart from the other wolves and wondered how he felt. 06 was nearly three, similar to a woman in her mid-twenties. The young males were not yet two, equivalent to teenage boys. 302 would soon be nine, over four times their age.

06 soon took off by herself, and the Blacktail wolves went back to their territory to the west. But 06 was not quite done with those males. Three days after 06 was bred by Medium Gray, I got a report that she had been spotted in Blacktail territory. One of the black males came to her, and within thirty seconds the two were in a mating tie. Soon four other Blacktail wolves charged in. The lead gray, who must have been 693, pounced on 06, but it was too late to prevent the mating. I wondered if 693 was being so aggressive because she wanted to prevent 06 from joining the Blacktail pack on a permanent basis.

That black male was Small Blaze, the fifth male 06 bred with late that winter. Apparently none of them impressed her much because she was still single at the end of the mating season. 06 seemed to be in no rush to pair off and settle down.

After that we saw 06 feeding on a fresh elk carcass, indicating she had made a solo kill. Over time we realized that she was a master hunter who could make a kill without help from any male. I once saw her confront an elk face to face, deftly dodge its charges and attempts to stomp her, then leap up and grab its throat. She maintained an iron grip until the elk collapsed and died. Perhaps her hunting prowess contributed to making her such an independent female. Few males were on her level, and she was doing just fine living the life of a lone wolf who seemed to have no need of a male partner.

During that mating season, we saw 302 breed only one female: 692. However, two female pups born into the Blacktail pack that spring were later collared, and the genetic analysis of one of them showed that 302 and 693 were her parents. That meant that 302 must have mated with 693. He probably bred 642 as well, who also had pups that spring. 302 was old, but he still was sought after by much younger females.

We saw more wolves with mange that winter. Some were missing fur just on their tails and legs, while others had lost much more of their coats. The skin in those bare patches often was black. A Wolf Project crew examined one dead wolf who had a severe case of mange: 90 percent of his coat was gone. His bone marrow was depleted and watery rather than solid. Marrow fat is usually the last reserve used by a wolf or other animal in poor condition.

I SAW THE thirteen wolves in the Druid pack on February 27. Last year's pups and the young adult females were playing and chasing each other and 480 could not resist coming over and joining in. The big alpha male wrestled with the much younger members of the pack. As he did not use his full strength, the matches seemed to be among equals. Pups would run off, come back, and run circles around him. 480 seemed to be pretending he was a pup or yearling. Wildlife behaviorist Marc Bekoff, in his 2002 book *Minding Animals: Awareness, Emotions, and Heart*, calls that role reversal or self-handicapping. In this case, a big alpha male was acting like he was a low-ranking pack member.

Marc writes that "social play might be a 'foundation of fairness' and can provide insights into the evolution of social morality." During the hundreds of times I have watched adults and pups play, it is obvious that they enjoy themselves, or to put it more simply, they are having fun. A pup that is bigger and stronger than its littermates could easily win every playful wrestling match and catch every pup in chasing games. But if they did that, the other pups would likely avoid playing with them. Perhaps if that big pup were in 480's family and saw how his father pretended to lose matches and chases, the pup would come to an understanding that equality or fairness in play sessions extends the enjoyment. The principle is simple: you chase me, then I chase you; I will let you outwrestle me, then I will win the next match. Marc classifies that as reciprocity in play.

I would think that the concept of fairness learned by young pups likely goes on to facilitate cooperation among adult wolves when hunting and sharing food with pack members who have stayed behind at the den to watch over the pack's pups.

In Marc's 2007 book *The Emotional Lives of Animals*, he uses a quote from Charles Darwin's groundbreaking 1871 work, *The Descent of Man, and Selection in Relation to Sex*, that captures this principle of fairness and cooperation that I have seen in wolves: "Those communities which included the greatest number of the most sympathetic members would flourish best and rear the greatest number of offspring." Based on my experience watching wolves like 8, 21, 42, and 113 in Yellowstone, I would say that wolves who are the fairest and the most cooperative have the most allies, and those allies treat them in a reciprocal manner.

I saw the Blacktail wolves the morning following that big Druid play session. The younger adults led while the much older 302 lagged behind. When the lead pack members made brief stops, 302 would immediately lie down to rest. He would then get up and follow once the others moved on. Later they stopped for a longer period and alpha female 693 lay down next to 302. I got the sense that she liked to be near him.

By early March, there were more changes in the Blacktail pack membership. The gray male who had left earlier now seemed to be part of the Agate pack. He had been joined by his brother Big Blaze, who had become the Agate alpha male. They were the only two males in the Agate pack. 472 was still in the group, along with 07 and 692, the black female who had helped form the Blacktail pack but later returned to her mother and original family. 692 had been bred by 302 while she was with the Blacktails and looked pregnant. 472, who had bred with both of the Blacktail males now in her pack, also showed signs of pregnancy. All the complicated romantic activity that season among the wolves made me feel like I was watching a television soap opera.

471, a gray female and a relative to 06, had previously left the Agates and started a new pack between Mammoth and Blacktail Plateau. It was known as the Lava Creek pack, and a black wolf, 147, was the alpha male. 06 was in that group off and on as well, but she continued to spend most of her time as a lone wolf. 147 had patches of bare leathery skin on his hindquarters, an indication that he had suffered from a bad case of mange but was now recovering.

There was not much we could do to stop mange from spreading. Jon Way, a coyote researcher, told me that veterinarians treat dogs with mange by giving them a pill on three separate occasions, two weeks apart. That would work fine on a pet dog but could not be done on a wolf living in the wild. Biologists have experimented with dropping baits with pills hidden inside from planes flying over areas where coyotes have mange. That can help the animals recover. However, if a coyote eats too many pills, the medicine could kill off the good bacteria in its digestive system.

The departure of Big Blaze and 692 left six wolves in the Blacktail pack: females 693 and 642, and males 302, Big Brown, Medium Gray, and Small Blaze. The Wolf Project staff and the people who had known 302 for years were especially interested in how he was going to function now that the denning season was approaching. He would soon be weighed down with heavy responsibilities such as providing food to the mother wolves and the soon-to-be-born pups. The success or failure of the denning season was going to rest primarily on his shoulders.

23

302 and His Pups

THE BLACKTAIL WINTER Study team told me an intriguing story about 302. They had seen him bedded down with the three male yearlings in his pack in March. When pregnant alpha female 693 arrived, all four males ignored her. She went to 302 and pestered him for a regurgitation. When he did not respond, she sat on his head. That got him up, but he continued to ignore her. Later younger female 642, who also was pregnant, came in. 302 and the other males gave her a big greeting. Just as those males had preferred 06 to 693, they now seemed to like 642 more than 693. I do not have an explanation for that difference in response other than that, like people and dogs, wolves seem to have preferences for who they like to be with.

Rebecca Raymond, who was on the Blacktail Winter Study team, told me another 302 story. After he had been bedded down for some time, he got up, picked up a big piece of meat that he had hidden under his chest, carried it over to 642, and gave it to her. That was just how an alpha male should take care of a pregnant female.

By early April, I was often getting Druid signals at their traditional den, a sign they were intending to den there once again. At times alpha female 569's signal got very weak, then I would lose it, indicating she had gone underground.

On April 9, I saw the gray male in the Agate pack, the former Druid and one-time Blacktail wolf, with part of an elk hide. He plucked fur from it, then ate the hide. That indicated the pack had not made a kill recently and were so hungry they had resorted to scavenging an old carcass. 692 came over and also ate a section of hide. Then both wolves grabbed the same section of hide and had a tug-of-war to determine who would get to eat it. Soon after that, I saw some of the Druid adults scavenging on an elk that had died the previous winter, gnawing on bones scattered around the site. Those sightings concerned me because they indicated the Agates and Druids were not eating well right when the mother wolves needed plenty of food to support their litters.

A shortage of better-quality food may have been what caused the previous year's pups to scavenge whatever they could find on their own. One morning a Druid pup dug into a snowfield and came back up with a dead vole in its mouth. Saving it for later, the pup placed the tiny rodent off to the side, then resumed digging for another vole. That was a mistake, for a nearby raven hopped over, grabbed the vole, and flew off with it. The pup got another vole but this time ate it right away.

I saw the first bison calf of the spring on April 11. Bison calves are red when born, and local people call them "red dogs." The newborn figured out how to get up on its hind legs but could not yet manage standing on its front legs. The start of the bison calving season was a good development for the wolves for they would be able to scavenge on stillborn calves and get sickly or poorly defended ones.

302 and the three other males in his pack were often seen near the original Leopold den site in the Blacktail area, the den where 302 had been born. That suggested that at least one of the Blacktail females was now denning there—we later determined that it was 642. I checked on 693 and got her signal from a den located a mile or so to the east. I also picked up 302's signal from that direction. Despite his disinterest in 693 back in March, he was now dutifully attending her. Meanwhile the younger males were with 642. As I continued to check on the wolves, 302's signals indicated he was regularly helping the mother wolves at both sites.

In mid-April, twelve inches of snow fell in the Blacktail area. I put on snowshoes and trudged up to the top of South Butte where I could look at 302's territory. The three young males—Big Brown, Medium Gray, and Small Blaze—were at 642's den. That day two of 642's older relatives visited her: 471 from the Lava Creek pack to the west and 692 from the Agate pack to the east. I found it fascinating that females from neighboring packs had the right of free passage to visit their relatives' dens. I had seen similar behavior among related females during earlier denning seasons when territorial rivalries seemed to be suspended and wolves from neighboring packs came to check out how their former pack-mates were doing now that they had families of their own.

Early on April 18, I saw twelve Mollie's wolves near the Druid den. Two adults had bad cases of mange and a yearling had a less severe case. I worried that they might spread the mites to the Druids. So far there was no sign of mange on any of them.

I was getting signals from Druid alpha female 569 and a younger female at the den. 480 was not in the area. Had something happened between the two rival packs during the night? If 569 was in her den, she could probably defend it by stationing herself just inside the entrance, like the Slough females had done in 2006, but she still needed to eat. The Mollie's wolves did not approach the Druid den. The next morning, 480 was back and the Mollie's wolves were across the road, several miles to the southwest. I watched as the Mollie's pack traveled farther south and disappeared over the top of Specimen Ridge, heading toward their territory.

I was more concerned a few days later when I saw that some of the young Mollie's wolves had returned and were socializing with several young Druid adults. A face-to-face greeting that included sniffing and licking could easily transfer mites from one wolf to another.

BY LATE APRIL, 692, who had been dividing her time between the Agates and the Blacktails, was mostly with the Blacktail wolves. Although she had been bred by 302, we thought that she was going to have her pups in Agate territory. But she abandoned her den there and reunited with her two sisters in 302's pack. One day I spotted 642, 692, and 693 with three of the Blacktail males at a fresh bison carcass. All three females had distended nipples, which established that they were lactating. 692 had apparently lost her litter, which might have been why she left the Agates, but she could now help out by nursing her sisters' pups. With 692 back with them, the Blacktail pack had seven adults.

Druid female Dull Bar had left Lamar Valley and started a new pack with another Druid female and a new male. But the group fell apart after another pack attacked them, and Dull Bar returned home alone. She appeared to be lactating and, like 692, could use her milk to feed her sisters' pups. Including Dull Bar, there were now fourteen adults in the Druid pack.

By that time, the Slough pack was no longer a threat to the Druids because there were only five wolves in the group: the former Agate alpha male, 383, and four females, three of whom had some level of mange. The group had abandoned the Slough Creek area and was usually seen west of there. The multiyear conflict between the Sloughs and Druids was over, with the Druids the winners, due partly to 480's leadership and partly to the Sloughs' run of bad luck.

The 06 Female continued to divide her time between the small Lava Creek pack and the Agates, her birth family. A younger Agate wolf, the 07 Female, often bullied and dominated 06, just as 693 had done. The bullying from the two females probably factored into 06's choice to live independently most of the time. She seemed content to live her life alone, at least for now. Despite having bred with several males, 06 apparently did not get pregnant that year.

WHILE I WAS monitoring the packs, I continued to give frequent impromptu talks in Lamar Valley for groups of visitors and for schoolchildren on field trips. I felt a responsibility to tell people about the successful story of wolf reintroduction in Yellowstone and the histories of the local packs. I especially liked to tell the stories of wolves I had followed and admired over the years—wolves like 8, 9, 21, 42, 480, and now 302. Prior to 2008, I did not keep records of how many of those talks I did, but I started to keep a list that year. I would give over eighteen hundred talks to visitors over the next ten years.

Wolves are not the only wildlife superstars in Lamar Valley. Two local men, Bill Hamlin and Ralph Neal, have spent a lot of time in the park over the years. They specialize in spotting grizzlies and usually find several every day, sometimes more than twenty. Bill and Ralph like to invite park visitors to see the grizzlies through their high-power spotting scopes. One day Ralph told me that he and Bill had found a mother grizzly with two new cubs and showed them to five hundred people.

The local bear watchers call first-year bears "cubs of the year." One day a new visitor heard someone refer to a mother bear and cub of the year and asked, "How does a bear get voted Cub of the Year?" Ralph and Bill are happy to give up their time at their own scopes and give priority to people who have never seen a grizzly family before. The wolf watchers have the same sharing custom and gladly step away from their scopes to let a new person see a wolf.

By 2009 Yellowstone was getting as many as four million visitors annually, which meant there were sometimes problems involving wolves and people. A wolf repeatedly approached park visitors that tourist season, and we got reports that it was because people had fed the animal.

222 | THE REDEMPTION OF WOLF 302

Rangers tried to change his behavior by hitting him with rubber bullets and using other forms of aversive conditioning, but none of it worked. The wolf appeared to be too addicted to human food to stop and had to be shot for public safety. That never would have happened if visitors had not fed him.

WE STARTED TO see newborn elk calves in late May. The pregnant mothers usually go off by themselves to give birth. When cows are scattered throughout a large region, it is harder for wolves to find a newborn. The mothers rejoin a herd when their calves are capable of keeping up with the group as they travel from one feeding area to another. Wolves can find calves much more easily then, but cows will team up to drive wolves away.

One morning I watched seven Druids spread out to search an area of thick sagebrush for elk calves. A cow elk was nearby, monitoring the wolves. She must have had a calf hiding nearby but did not do anything to give away its location. The Druids mostly ignored her and continued the search. Then I saw the wolves run into a gully. They must have seen a calf there. I went to a different angle and could see that the Druids' searching strategy had paid off: they had found a calf.

Kira Cassidy flew in late May and saw five pups playing together at 642's den site in the Blacktail area. The alpha pair, 302 and 693, were nearby. The surviving pups born at 693's den must have been brought to that more central location because she was based there from then on. A few days later, 302 was spotted feeding on a carcass. He then went directly back to the den to feed the pups. That was more proof that

302 had become a capable and responsible alpha male who was taking good care of his family.

On June 12, I completed my ninth year of going out early every day to study wolves. I hoped to keep on doing it for at least another year so I could make it ten years. If I could continue to fourteen and a half years, that would double Cal Ripken Jr.'s record of starting 2,632 consecutive games.

I saw 302 and 693 near a fresh carcass in the Tower Junction area later that morning. Doug Smith was doing a flight at the time, and he called down to say 642 was attending six pups at the Blacktail den. The pups were different sizes, likely because they were from two different mothers. Four pups were black and two were gray. The sighting of six pups turned out to be the high count.

My first sighting of pups at the Druid den was in mid-June. There were smaller and bigger ones, which indicated that 569 and at least one other female had litters. The high count of pups was nine: five blacks and four grays. Around that time, I noticed that a young gray Druid had mange and wondered if Light Gray had spread it to that wolf. I had dreaded that might happen and now it was reality.

I went to the Blacktail area in late June to check on 302 and his family and once again climbed South Butte. I had been watching wolves from that observation point since the spring of 1996 when the Leopold pack's founding alpha pair, male 2 and female 7, were raising their first set of pups. When I got in place, I spotted 302 walking around the area. Nine years after he was born there, 302 had come home and was raising pups in his parents' territory. Some of his pups had been birthed in the same den where he had been born. I wondered if 302 thought about his father, wolf 2, as he went about faithfully fulfilling the same duties 2 had the year 302 was a pup.

Dan Stahler did a flight around that time and told me about seeing the Quadrant pack in Gardner's Hole, a valley southwest of Mammoth. When the Blacktails hunted in the western end of their territory, they sometimes strayed into the adjacent Quadrant territory. There were seven wolves in that pack—four adults and three pups.

In early July, I went back up on South Butte and spotted three Blacktail pups chasing each other around a rock. Later one of them dropped down into the ambush position and charged at another pup when it went by. After that they played king-of-the-mountain: one pup got up on the rock and the other two tried to climb up and knock it off. The upper pup jumped off the rock, then jumped back up on it. Another pup grabbed its tail and yanked it to the ground. The two mother wolves, 642 and 693, were bedded down nearby as their pups played. In my talks, I often speak about how similar human social and family behavior is to wolf behavior. That scene of the wolf mothers and their pups could have easily taken place at a playground with two women watching their kids playing.

Later a black pup that looked like a miniature version of 302 came out of the trees and traveled throughout the area. It had an independent streak, probably inherited from its father. The pup looked fat and healthy, proof that the adults were taking good care of the six pups. The thought came to me that this was a well-run and well-functioning pack. It was everything a classic wolf pack should be.

A gray pup joined the black and the two went to the bedded 642 and greeted her. The black pup rolled on its back
next to the female and pawed at her face. Both pups then
went back into the trees. We got 302's signal from there, so
the pups were probably with him. The big male known as
Medium Gray approached the trees and the pups ran out and
mobbed him. They licked his muzzle, and he lowered his
head and regurgitated meat for them.

I was back there a few days later and saw two of the young
males and 693. She brought an elk antler to the pups to play
with. Then I spotted 302 bedded on a nearby hill, watching
over his family like a benevolent patriarch. Patriarch was an
appropriate word because his once jet-black coat was now
streaked with gray.

When a black pup stalked something I could not see, 302
stood up to monitor it, like he wanted to be sure it was safe.
The pup suddenly ran forward and pounced on something
small, probably a vole. The pup shook its head vigorously,
then ate the little animal. The little wolf was on its way to
being a master hunter.

On July 9, I saw two Blacktail pups independently exploring areas well beyond the den region. The three adult females
and other pups were near the den. They had a group howl
and both pups howled back. That suggested that the young
pups already understood the long-distance communication
system used by wolf families. I later got a good look at the
coats of the Blacktail adults and was relieved to see that all of
them were free of mange.

One morning 693 went to the marsh and met up with
the six pups, and they followed her behind a hill. Soon 302

came in, probably from a carcass, and looked around for the pups. He howled but did not get an answer. The wolf found the scent trail of the pups and followed it to where I had lost them behind the hill. He likely fed them there. After that I thought about how the Leopold pack no longer existed. Of all the sons born to the original Leopold alphas, it was 302, the least likely one, who came home to his parents' homeland, reestablished a pack there, and was raising pups who were the grandchildren of the founding alpha pair.

By late July, the ground squirrels were getting fat enough to start hibernation. I saw a gray wolf in Lamar Valley going back and forth with his head down, hunting for them. He grabbed a big one and pulled it into his mouth like a strand of spaghetti, then chased another squirrel, caught it, and vacuumed it up.

THE BLACKTAIL WOLVES had figured out something about crossing the road that the Leopold pack never did. When wolves in that territory wanted to cross the road, they had to go through open country where they would be easily seen by people driving through the area. Many drivers would speed to the likely spot where the wolves would cross and stop in the middle of the road. That often caused the wolves to back off and try to cross elsewhere. I got a report that the Blacktails had found a place where the road crossed a small bridge over a creek. The pack took to traveling under that bridge to the other side of the road, completely avoiding any traffic. I guessed that 302 came up with that innovation. I knew he was smart. Maybe his many years of sneaking into the territories of other packs to breed their females had helped him figure out routes where he could travel undetected.

I drove back to the Blacktail area on the last day of July and heard that six adults, all but 302, had earlier gone to a new carcass north of the road. I spotted 302, who had apparently stayed behind to watch over the pups, to the south. This was a job 302 was increasingly taking on as he aged, possibly because he was finding it difficult to keep up with the much younger adult members of the pack. He was slipping into a grandfatherly role, just as alpha 113 had done in the Agate pack in his old age.

In early August, I was back up on South Butte with Rebecca Raymond. Females 642 and 693 and the six Blacktail pups were in sight, and 302 was bedded down about a half mile away. Rebecca had gotten in position before me and told me that when 302 arrived from a carcass, two pups and 693 ran to him. He regurgitated meat for them three times. Two more pups joined the group and all four pups got another feeding. Later 302 got up, searched for the two pups who had not joined him, and greeted them. Rebecca then saw 302 bed down with a pup, licking and grooming it for some time.

Like all healthy young animals, the pups were curious about their surroundings and seemingly fearless. A while later, a black pup spotted a great blue heron. Blue herons are big birds, up to four and a half feet tall with a six-and-a-half-foot wingspan. The heron had already seen the little wolf and was watching it. The pup moved toward the bird and it walked away, then flew up and glided a few feet. When the heron landed, the pup continued toward it and the bird flew away once more. Two pups then pursued the bird and, on its next flight, the heron flew much farther off. The pups gave up at that point, probably after learning the futility of trying to catch such a big bird.

I went back to the Blacktail territory one mid-August evening. 302 was bedded down on a hill, and all six pups and four of the adults were scattered around the area. The pups played together and 302 monitored them from his hill as they ran around. A pup went to Big Brown and harassed him and he gave it several holding bites. They must have been gentle for they did not deter the pup. The large male got up and walked off, then bedded down again, probably hoping that would give him some peace. But the pup ran over and pawed at his face. Big Brown walked farther away and lay down again. When two pups harassed him, he moved off for a third time. Giving up on him, the pups ran around and played together.

I looked over at 302 and saw that he was studiously watching them. The pups were likely in an area where he had played with his littermates when he was a pup nine years earlier. Later three pups tried to sneak up on 302 from behind but he must have heard them because he turned his head. The pups wagged their tails at him, then moved on, leaving the old wolf to rest in peace.

When the adults and pups later traveled north, 302 was last in line, slowed down by a limp. At times he was three hundred yards behind the others. The lead four adults soon bedded down, and it looked like they were waiting for 302 to catch up. 302 seemed tired and acted like his leg was painful. I did not know the cause of his injury but suspected he had been kicked by an elk during a hunt.

A few mornings later, I saw 302 bedded down in a marsh that the Blacktails often used as a napping spot in hot weather to keep cool. Mother wolf 642 went to 302, greeted him, then took off and traveled throughout the area, looking for the six pups. I had not seen them anywhere. 642 went to spots where the pups had been recently, looked around and howled, then moved on to other sites. She howled periodically, but apparently never heard the pups answering her. She ran off to continue her search to the northwest.

I looked around and saw that 302 had gotten up and was going north. Then I spotted five of the pups a half mile from where I had last seen 642. 302 was headed straight for them. The hills between 302 and the pups would have prevented him from seeing them. The pups had not howled, so he could not have gotten the right direction by hearing them call out. He did not seem to be following a scent trail. I could not figure out how 302 knew where the pups were.

I soon saw that 302 was getting close to the pups. One of them spotted him and ran toward him. The other pups followed, also at a run. The lead pup reached 302 and jumped up to greet him by licking him in the face. The other pups soon arrived and all of them joyfully jumped around their father. He good-naturedly tolerated their excited greetings.

302 turned around and went back the way he came. The pups ran after him and, to reward them, he regurgitated meat for them. They fed at the site, then followed 302 single file when he continued south. At times 302 looked back over his shoulder to make sure they were still behind him. The scene looked like young soldiers marching behind a veteran sergeant.

He arrived at the marsh, the same site he had been resting in earlier, and lay down in a thick willow patch. The pups ran into the marsh, looked around for 302, did not see him, then spotted one of the gray males and rushed over to him. He acted annoyed and snapped at them. When the big male walked off, the pups followed.

AS I THOUGHT over the sequence of events, I was very impressed with 302. The mother wolf was frantically trying to find the pups but failed to get responses to her howls and could not find their scent trail. She ran off in the wrong direction. Then 302 sacrificed his nap time, went directly to them like he knew all along where they were, and led them back to the rendezvous site. I wondered if he had deliberately taken them to where the gray male was in the marsh, then hid in the willows so the pups would harass the gray, rather than him. That would free up 302 so he could take a long nap. It was a clever trick I would not put past him.

When he was a pup in this area, 302 would have known all the places that he and the other pups liked to explore and probably never forgot those sites. On seeing 642 frantically trying to find the pups, he got up, apparently made a good guess where the pups would be, found them, and led them back. He made it look easy.

The Druids had been based up the Lamar River for a while and had spent most of their time out of sight. I had spotted the alphas and four other adults in early August and had noticed something that concerned me: fur was missing in the middle section of 480's tail—probably due to mange. The end of the tail was fully furred, so it looked like a poodle's with a pom-pom on the end, not a good image for a tough alpha male. I also saw that a gray female had little fur on the lower third of her tail. The other wolves had no sign of mange. The young Druid I had noticed earlier with patches of mange was not in the group, so I could not see if he was recovering or getting worse.

Kira did a flight in mid-August and picked up Druid signals in the Cache Creek area but could not see them. She did spot the Mollie's wolves and circled over them. The ones with mange had fully recovered and all now had good-looking coats.

The Druids were back in the Chalcedony rendezvous site in late August. Doug McLaughlin was the first to spot them. I got there later and he pointed out wolves to me: twelve adults and three pups. The pups were all small and skinny. Two had thin areas of fur while the third one had hairless areas on her sides and tail. An adult black female had mange on her sides and legs. 480 and some of the adults that earlier had missing fur on their tails seemed better now. If the Druid pups stayed healthy and well fed, their coats should recover, but I was worried about how thin they were.

As I watched the pack, I saw that the pups were very active, which was a good sign. One pup went to the marsh and hunted for insects and voles, and the next morning the adults did a lot of regurgitations to the pups. A few days later, there were four pups at the rendezvous site. The new one was also skinny, and the fur on her sides was thin. It would be a while before we knew if the Druids, like the Mollie's wolves, would make a full recovery. 302's pack, the Blacktails, were, as yet, unaffected by mange, but in a few months' time they would have to contend with a very different threat, one that would test 302's mettle as an alpha to the limit.

No Country for an Old Wolf

D AN STAHLER FLEW on September 1 and saw 302 and five other Blacktail adults slightly east of the park road that runs south from Mammoth. They were about six miles from Mammoth, near Swan Lake, which put them at the western end of the Blacktail territory. The four adults in the Quadrant pack were on a carcass nearby. The two packs were close to each other but on opposite sides of the park road. I figured that the Quadrants regarded the area west of the road as their territory and the Blacktails felt the east side of the road was their land.

In two weeks, a legal wolf-hunting season would open in Montana. As there was now a thriving wolf population in the northern Rockies, wolves in that region had been taken off the federal endangered species list. Except for national parks, wolf management had been transferred to the local state fish and game agencies, which had the legal right to institute wolf hunting. There were four Montana hunting districts adjacent to the park and each had a limit of twelve wolves that could be shot. The park's northern border was also roughly the northern boundary of the Blacktail territory, but wolves do not always stay within their pack's territorial limits, especially when food is hard to find. Those of us that had known 302 for so many years hoped that he and his family would stay south of that border.

I talked to Wolf Project staff who were monitoring the Blacktail wolves and heard that the adults were not having much luck in hunting elk in the central part of their territory. They had to travel far and wide to find animals they could catch and take down. 302 and the other adults were often away from the pups for up to two days at a time. The pups knew how to hunt for small prey and survived on voles and insects while waiting for the adults to return. With the Quadrant wolves to the west and the wolf hunters to the north, I began to get concerned about the pack seeking out elk in hostile areas.

I did not get any signals from the Blacktail wolves when I did a check in the central part of their territory on September 19, so I continued toward Swan Lake. I pulled off the road south of Mammoth, took out my telemetry equipment, and picked up signals from the Quadrant wolves to the southwest. I drove a few miles farther south and got signals from 302 and 693 to the west. That meant they had crossed into Quadrant territory.

John and Mary Theberge were doing research on wolf howling in the park at the time, and we hiked out together to look for 302 and his family. When we reached a big open meadow, I got signals from the Blacktails to the southeast. The Quadrants were now to the northwest. There was no indication the packs were aware of how close they were to each other.

About forty-five minutes later, we spotted the Blacktail wolves coming out of the trees that bordered the meadow: 302, five other adults, and five pups. His limp seemed less pronounced. 302 was leading west, farther into Quadrant territory. He stopped to howl. I did not pick up any answering howls from the Quadrants but 302 must have heard something, for he immediately led his group south, away from the other pack, exactly the right strategy. As he traveled, he frequently looked back at his pack members, like he was making sure they were following him out of the area. Then he veered southeast into trees and I lost the group. If they kept going in that direction, they would reach the park road. When they crossed it, the Blacktails would be safely back in their territory.

As far as we knew, the Blacktail wolves had never run into the Quadrants, and 302 would not know if his pack was bigger or smaller. Leaving was the prudent thing for him to do. The area where we found him that day was about nine miles west of the Blacktail den and rendezvous site. There were usually a lot of elk in this part of the Quadrant territory, so that must have been the attraction for 302 and the other adults with him.

The next morning, I drove back to the same spot and picked up signals from 302 as I approached Swan Lake. I heard that wolves had been seen nearby, east of the road. Soon I heard howling from west of the road, the direction where 302's signal was coming from. Then I saw four adults to the east, in Blacktail territory, looking across the road in the direction of 302's howls. Those wolves moved toward a nearby area with a lot of ravens, a sign the Blacktails had made a kill there. No pups were with them, so they were probably in Quadrant territory with 302. I did not get Quadrant signals, which meant they did not know the Blacktails were in the area.

Soon after that, I saw 302 with two Blacktail adults west of the road. The two adults crossed the road and headed toward the Blacktail carcass site. Then 693 crossed in the other direction, going toward 302. I took this as more evidence that the pups were with 302. I thought the two adults who had crossed the road would likely feed at the carcass and head back to regurgitate meat for the pups.

I was worried. I did not know why 302 and the pups were lingering in Quadrant territory. What would happen if the Quadrants detected the intruders and charged at them? It was not a good situation, especially for 302, who was having trouble keeping up with the rest of his pack. This was no country for an old wolf.

Five of the pups were safely back at the Blacktail rendezvous site on September 25. Rebecca Raymond drove to Swan Lake and located all seven adults, including 302, in Quadrant territory just west of the road. Those incursions into the rival pack's territory were risky, but so far, the two packs had not run into each other.

On October 5, I saw all the Blacktail adults and five pups well out of Blacktail territory, a mile west of Slough Creek. By that time, nine Yellowstone wolves had been shot north of the park border, which was just five miles from the Blacktails' current location. The quota in that area was twelve, so three more could be taken.

I spotted twelve Blacktails north of Mammoth, just inside the park's North Entrance, the next day. That area was a straight-line distance of seventeen miles from where I had seen the pack the previous day, but since wolves do not travel in straight lines, the true distance would have been much longer. Seven adults and five pups were in the group. We concluded that the sixth pup, a gray, had been lost to unknown causes. If the pack had gone north another mile or so, they would have crossed the park border and entered the wolf-hunting zone, but by the time I spotted them, they were heading back south. That relieved my worries about the family being shot, but by going south they were heading toward the Quadrant wolves, which put them at risk of getting into a battle with a rival pack.

I drove south a few miles as I followed their signals and saw the pack across the Gardner River, still headed south. 693 was leading and 302 was well behind. The Blacktails were traveling through an eroded canyon and had already passed through numerous steep ravines. 302 was having a hard time getting back up those gullies. At one point, he stopped to rest as the pack moved on. When 302 got up, I could see that he was not doing well. The long trip from the Slough Creek area in the past day must have exhausted him. The rest of the wolves, all many years younger than he was, wanted to continue traveling and he was struggling to keep up.

That reminded me of a sighting Rebecca had mentioned to me. The Blacktail wolves were on a long hunt and 302 was lagging behind. He bedded down to rest as the others continued without him. 302 howled as he lay there, all alone. The pack made a kill and alpha female 693 later came back, found 302, and took him to the carcass. Her story was touching for it showed the strong emotional bond 693 had with 302.

As the pack continued traveling away from the park border, they crossed the park road to the south, a mile east of Mammoth, and ended up in the Swan Lake area. I thought about the long distance 302 had traveled from Slough Creek to here as he struggled to keep up with his pack. I called that trip 302's Long March. Kira Cassidy later checked our GPS data on 302 and found that he had started that trip on October 3 near Swan Lake. From there to Slough Creek and back to Swan Lake was about fifty miles.

The sighting that day and Rebecca's story concerned me. I realized that I thought of 302 as being forever young. The reality was that he was nine and a half years old, about seventy-four in human years. I worried that if his strength declined much more, he would not be able to keep up with the pack. Then I realized there was another, more positive, way to look at 302's behavior that morning. Despite his advancing age, he had not given up. He chose to keep following his family. I thought again of 113, the founding alpha male of the Agate pack. A combination of old age and a serious injury had taken a heavy toll on him. Like 302 he also had trouble keeping up with his pack. 113 disappeared one day and, because his radio-collar battery was dead, we could not find his body. Years later a research crew found his collar miles from his family's territory, indicating he had taken off by himself and almost certainly died alone.

I also remembered how at the end of his life, 21 left his family to die alone in a high-elevation meadow where he and 42 had spent so much time together. Whenever 302's last hour of life came, I hoped that some of the members of his family would be with him.

ON OCTOBER 7, I got only one weak signal from a Druid wolf in Lamar Valley. Deciding it would not be worth it to look for the Druids that day, I continued west. I heard that Kira had gone south from Mammoth and picked up signals from 302 and other Blacktail wolves about six miles down the road. She also got signals from the Quadrant alpha male to the west of that area.

I drove there, stopped at a service road south of Swan Lake, and got loud signals from 302. A man there told me he had seen ravens flying off with meat in their beaks. I walked around and spotted fresh wolf tracks in the snow close to where I had parked. I did a direction check on 302 and he seemed to be to the south, probably just west of the main road which put him in Quadrant territory. Ravens were loudly calling out from there, so they were likely waiting to feed at a carcass. 302 had to be at that site.

I got a call from a wildlife guide and he told me he had just seen five adult wolves a few miles south of where I was parked. I drove there and saw 693 and other adult Blacktails west of the road. I lost them going south. The group was traveling away from 302's location. I felt that meant the carcass was mostly consumed. They were off on a new hunt, and 302 had stayed behind, probably because he was too tired from the pack's recent long march. There were no pups in that traveling group, so they had likely stayed behind with 302. If he could not keep up with the hunting party, he could serve the pack by taking care of the pups.

Normally, if just one adult was taking care of the pups in the middle of the Blacktail territory, it would not be a dangerous duty because a rival pack would not likely attack them there. But 302 and the pups were in an area claimed by the Quadrant wolves, and the carcass might originally have been theirs. Even if it was not theirs, they could easily detect it by spotting the ravens coming and going from the site or by picking up the scent of the dead animal on the wind. I envisioned a scene where the Quadrant adults ran into the site, saw 302 and the pups, and attacked the wolves who were invading their territory. I drove back to the north and still got 302's signal from that area. I did not get any Quadrant signals, so that seemed to mean 302 and the pups were safe. I drove off to check on other wolves back to the east.

It was 13 degrees Fahrenheit (–11 Celsius) when I left my cabin at 6:20 a.m. on October 8, and new snow was on the ground. I heard on the park radio that the road south of Mammoth was closed because of heavy snow, but it had opened by the time I reached the area. I turned down the road, and about fifteen minutes later I pulled into a service road south of Swan Lake and checked for 302's signal. I got a loud transmission, but the beeps were twice the normal rate. That was the mortality signal, indicating the motion detector in the collar had not registered any movement for four hours. That could mean one of three things: his collar had fallen off, the electronics were malfunctioning, or 302 was dead.

I called Dan at the office and filled him in. He put together a team to come out to my location. While I waited, I drove south and spotted 692 and three pups east of the road. I then heard howling farther to the east that seemed to be from additional adults. It looked like 692 and the pups were heading that way. Another adult wolf appeared, also traveling in the direction of the howls. That accounted for two adults and three pups, along with what seemed to be several other howling Blacktail wolves.

I returned north to the service road and waited. Four people from the office soon arrived: Dan, Erin Albers, Kira, and Josh Irving. Erin operated the telemetry equipment as we walked out in the snow to search for 302. The terrain was flat, which made it hard to pin down the direction the signal was coming from. The five of us moved back and forth in a line looking for him or his dropped collar, hoping all we would find would be a dropped collar. But I admitted to myself that it was much more likely 302 was dead.

After forty minutes, Erin called out saying she had found him. We rushed over and saw 302 lying on his side by Erin's feet, just west of the road. At first we wondered if he had been hit by a car, but then we saw bite marks all over his body. Despite the injuries, my impression was that he looked peaceful, like he had died a good death.

His sleek black coat was in prime condition and his teeth were in good shape for an old wolf. I touched his torso and it was solid and muscular. I estimated he weighed about 130 pounds, big for a Yellowstone wolf. There was no sign of a fight or struggle at the site where we found him. The snow under him had melted from his body warmth. We saw a lot of wolf tracks surrounding his body. They were pup sized.

I figured that 302 and the six-month-old pups had gone back to the carcass in Quadrant territory sometime during the night and fed. The Quadrants, a pack of seven wolves with two big adult males, must have run into the area and chased the Blacktail wolves east, toward the road. Based on how I had seen 302 lagging behind the other Druids a few days earlier, he likely was last in line. But then a different thought came to me. I had seen 21 position himself between his pups and rival wolves, meaning that the other wolves would have to go through him to get to the pups. 302 had changed so much in his old age that I could see him doing the same thing. At some point between the carcass and the road, the Quadrants must have caught up with him and mortally wounded him. He managed to get as far as this spot before he lay down.

We picked him up gently, carried him back to the service road, and slid him into the back of a pickup. Word had spread about his death. Many people arrived and asked if they could see him. We let everyone come over and touch him if they wanted. Most of them had known and loved 302 and were crying. He had unknowingly deeply touched the hearts of countless Yellowstone visitors and millions of people around the world who had seen the documentary on his life. Once everyone had a chance to have a moment with the wolf, the crew drove him to park headquarters and did a full necropsy.

I got a strong signal from 693 as I drove off. We later saw the six Blacktail adults and all five pups, none of which showed any signs of injuries. The only wolf in the family that was harmed was 302. Big Brown took over the alpha male position, and the pack carried on without its famous founder.

When I was driving home, I had time to think about why so many people loved 302. I decided it was because he was so relatable. 302 was a flawed character who often let others around him down. Wolf 21 was sometimes called a perfect wolf because he selflessly devoted himself to his family and always seemed to do the right thing. 302 was an imperfect wolf, and that is what made him relatable. We are also imperfect, so he was one of us. The value of his life story is that in his final years he changed. We love stories about flawed individuals who turn their lives around and become heroes, for they give us hope that we also can choose to do better.

Dan later told me about the necropsy results. 302 was in good health for an old wolf. He weighed only 100 pounds, which was much lower than my estimate. He had lost a lot of blood, and that accounted for part of the difference, but not all. Even in death, 302 was larger than life. He had probably lost weight after traveling far and wide as he worked hard to feed his pups. I also thought that he likely shared much of his food with the pups, forgoing a full belly so they could have more of a share.

302's neck and throat had been penetrated by many wolf bites, indicating that several wolves had ganged up on him. There were also bites on his face, hips, and undersides. The bites to the face indicated that he had been fighting back against his seven opponents as best he could. Death was due to blood loss and shock. His belly was full of fresh elk meat, backing up our theory that he had been at a nearby carcass.

Genetic samples from 302 were collected and a complete genome sequence was done from that material, the most detailed genetic sequence ever done on a wild wolf. Later the same sequencing was completed on five of his sons and daughters. That research information will be used in DNA studies for decades to come.

IN THE FOLLOWING days I caught myself absentmindedly checking for 302's signals several times. His death had not fully sunk in yet. I thought about how it was a different world now that he was gone. I had known him since January of 2003, nearly seven years, and I had seen him go from a renegade wolf who refused to follow the rules to a dutiful alpha male who was devoted to his family. Yellowstone seemed diminished without him.

The main lesson I drew from 302's story is how he changed toward the end of his life. He had become a wolf like his father, wolf 2; his uncle, 21; and his nephew, 480. Being exposed to the behavior and role modeling of those accomplished males showed 302 how a wolf should act when confronted with difficult and dangerous situations and, in the end, he impressed us all by stepping up to that challenge.

For those alpha males, you could sum up their code of behavior in one line: if your family is hungry, if your family is threatened, you are supposed to do something about it, even if it puts you at great risk. 302 had failed those around him many times when he was younger, then managed to overcome whatever it was in his makeup or history that led to his failures. If 302 could do that, anyone could.

I thought a lot about the five Blacktail pups and how we had seen their tracks in the snow around 302. The melting snow around his body indicated that 302 must have been alive when he lay down after the attack. I pictured him during his final minutes of life seeing the pups running in. They would have greeted 302 with wagging tails and licked his face. Some likely also licked his many wounds.

That would have comforted him, but there was something else, something more important about the arrival of the pups: 302 would have seen that they were alive. 302's fight with the Quadrant wolves cost him his life, but however long that battle took, it gave those pups enough time to run away. He bore the full brunt of the attack and none of his pups were harmed. 302 was a father who died knowing he had saved his young sons and daughters. For an alpha male wolf, there is no greater accomplishment than that.

Later another thought came to me about the symmetry of the long-running story I had witnessed. 21 and 302 had a very contentious relationship. I had seen 21 catch his young nephew over six years earlier and beat him up. He could have fatally wounded 302 but let him go. Despite how 302 would get 21's daughters pregnant, then abandon them, 21 never killed him. Long after the death of 21, wolf 302 finally settled down and established a pack with three of 21's granddaughters. The pups that 302 saved that day were 21's great-grandsons and great-granddaughters. 21's long-ago choice to let his wayward nephew live led years later to that same wolf saving the lives of 21's descendants.

In Yellowstone there is a pantheon of the great alpha wolves. That includes females such as wolf 9 and wolf 42 and males such as wolf 8, wolf 21, and wolf 480. It took him a long time, but at the end of his life an unlikely individual redeemed himself and earned a place on that distinguished list: wolf 302.

EPILOGUE

———

WHEN 302 DIED, the 06 Female was three and a half
years old, middle aged for a wild wolf. She had had
many male suitors over the years but never settled
down with any of them. 06 seemed content to live life mostly
as a lone wolf, choosing to be on her own, living a free and
independent existence.

Then she met someone that changed her life, a male less
than half her age. To us he seemed much less impressive than
the older and bigger males we had seen court her. Whatever
06 saw in him, it caused her to radically restructure her life
and settle down. Soon that male's brother joined them and
06 became the leader of a three-member pack.

She was a granddaughter of 21. In the coming years she
would rival his accomplishments and have as great an impact
on Yellowstone and the people that came to see the park
wolves as her grandfather had. 06's story, and the stories of
other heroic female wolves in Yellowstone, will be told in my
next book.

AUTHOR'S NOTE

———

INSPIRED BY HOW members of a wolf pack support each other in time of need and by all the kind people who have helped me over the years, I will be donating proceeds from my wolf books to Yellowstone National Park and to nonprofit organizations such as the Make-A-Wish Foundation of America and the American Red Cross. Readers who are interested in helping support wolf research and wolf education in Yellowstone can go to the Yellowstone Forever website, www.yellowstone.org/wolf-project, to make a donation. Yellowstone Forever is the official nonprofit partner of Yellowstone National Park and helps fund the park's Wolf Project, the National Park Service operation I used to work for.

ACKNOWLEDGMENTS

I FIRST WANT TO thank my editor, Jane Billinghurst, for working way beyond the call of duty to make my original manuscript far more readable and polished than it would have been without her help. Thanks also to Rob Sanders at Greystone Books for accepting my proposal for this series. Everyone else at Greystone was very supportive and encouraging and thanks to all of them. Thanks to Rhonda Kronyk for doing the copyediting and to Meg Yamamoto for proofreading the final version of the book. I also want to thank Fiona Siu and Nayeli Jimenez for designing the look of the book and all the hardworking people who helped get it out onto bookstore shelves and into the hands of readers. Megan Jones has done a great job in promoting and publicizing my books.

My good friend Laurie Lyman read the first draft of the book and gave me very helpful comments and suggestions on how I could improve it.

There were many current and former National Park Service staff members and wildlife researchers who advised me on their work and experiences with wolves. Here are the ones who were especially helpful: Emily Almberg, Norm Bishop, Ellen Brandell, Jim Halfpenny, Tim Hudson, Josh Irving,

John Kerr, Bob Landis, Scott Laursen, Matt Metz, Abby Nelson, Rolf Peterson, Ray Rathmell, Rebecca Raymond, Doug Smith, Dan Stahler, Erin Stahler, Jeremy Sunder-Raj, John and Mary Theberge, Sue Ware, Jon Way, and Bob Wayne. Scores of volunteers have worked for the Wolf Project over the years and every one of them was very helpful to me.

I want to give special thanks to Kira Cassidy, who drew and illustrated the maps for the book. Thanks also to the photographers who allowed me to use their pictures in the book.

Special appreciation goes to Doug Smith, the lead biologist for Yellowstone National Park's Wolf Project. Doug is a unique PhD-level scientist who can relate what he has been learning about wolves and the natural world to regular people in ways that not only educate but, more importantly, inspire them.

There have been hundreds of wolf watchers who have greatly aided me over the years. On many occasions when I was trying to find wolves, another person would spot them first and graciously point them out to me. I would also like to thank the vast numbers of Yellowstone visitors who have been kind and friendly to me over my many years here. There is something about being in the park that seems to make people positive and sharing. Thank you to everyone I have met over the years. I could not have done this without all of you. I regard this book as a joint effort.

REFERENCES
AND SUGGESTED READING

Almberg, Emily S., Paul C. Cross, Peter J. Hudson, Andrew P. Dobson, Douglas W. Smith, and Daniel R. Stahler. 2016. "Infectious diseases in wolves in Yellowstone." *Yellowstone Science* 24(1): 47–49.

Bekoff, Marc. 2002. *Minding Animals: Awareness, Emotions, and Heart*. Oxford: Oxford University Press.

———. 2007. *The Emotional Lives of Animals*. Novato, CA: New World Library.

Bingham, Clara. 2016. *Witness to the Revolution*. New York: Random House.

Cassidy, Kira A., and Richard T. McIntyre. 2016. "Do gray wolves (*Canis lupus*) support pack mates during aggressive inter-pack interactions?" *Animal Cognition* 19: 1–9.

Cassidy, Kira A., Douglas W. Smith, L. David Mech, Daniel R. Stahler, and Matthew C. Metz. 2016. "Territoriality and inter-pack aggression in gray wolves: Shaping a social carnivore's life history." *Yellowstone Science* 24(1): 37–42.

Darwin, Charles. 1871. *The Descent of Man, and Selection in Relation to Sex*. London: John Murray.

Haber, Gordon. 1977. "Socio-ecological dynamics of wolves and prey in a subarctic ecosystem." PhD thesis, University of British Columbia.

Halfpenny, James. 2012. *Charting Yellowstone Wolves: A Record of Wolf Restoration*. Gardiner, MT: A Naturalist's World.

Haugen, H. S. 1987. "Den site behavior, summer diet, and skull injuries of wolves in Alaska." Master's thesis, University of Alaska, Fairbanks.

Landis, Bob, director. 2010. *The Rise of Black Wolf*. Trailwood Films.

MacNulty, Daniel R., Douglas W. Smith, L. David Mech, John Andrew Vucetich, and Craig Packer. 2012. "Nonlinear effects of group size on the success of wolves hunting elk." *Behavioral Ecology* 23: 75–82.

McIntyre, Rick, editor. 1995. *War Against the Wolf: America's Campaign to Exterminate the Wolf*. Stillwater, MN: Voyageur Press.

McIntyre, Richard, J. B. Theberge, M. T. Theberge, and Douglas W. Smith. 2017. "Behavioral and ecological implications of seasonal variation in the frequency of daytime howling by Yellowstone wolves." *Journal of Mammalogy* 98: 827–834.

Mech, L. David, Douglas W. Smith, and Daniel R. MacNulty. 2015. *Wolves on the Hunt: The Behavior of Wolves Hunting Wild Prey*. Chicago: University of Chicago Press.

PBS. 2007. *In the Valley of the Wolves*. Nature series.

Phillips, Jeanne. 2008. "Young Brother Gets Spotlight," Dear Abby column, *Billings Gazette*, June 17, 2008.

Smith, Douglas W., and Emily S. Almberg. 2007. "Wolf diseases in Yellowstone National Park." *Yellowstone Science* 15(2): 17–19.

Smith, Douglas W., Daniel R. Stahler, and Debra S. Guernsey. 2005. *Yellowstone Wolf Project: Annual Report, 2004*. Wyoming: National Park Service, Yellowstone Center for Resources, Yellowstone National Park.

———. 2006. *Yellowstone Wolf Project: Annual Report, 2005*. Wyoming: National Park Service, Yellowstone Center for Resources, Yellowstone National Park.

———. 2007. *Yellowstone Wolf Project: Annual Report, 2006*. Wyoming: National Park Service, Yellowstone Center for Resources, Yellowstone National Park.

Smith, Douglas W., Daniel R. Stahler, Debra S. Guernsey, et al. 2008. *Yellowstone Wolf Project: Annual Report, 2007*. Wyoming: National Park Service, Yellowstone Center for Resources, Yellowstone National Park.

Smith, Douglas W., Daniel R. Stahler, Erin Albers, et al. 2009. *Yellowstone Wolf Project: Annual Report, 2008*. Wyoming: National Park Service, Yellowstone Center for Resources, Yellowstone National Park.

———. 2010. *Yellowstone Wolf Project: Annual Report, 2009*. Wyoming: National Park Service, Yellowstone Center for Resources, Yellowstone National Park.

INDEX